495

M.

GENETIC STUDIES OF GENIUS

Edited by

Lewis M. Terman

VOLUME V
THE GIFTED GROUP AT MID-LIFE

Thirty-five Years' Follow-up of the Superior Child

GENETIC STUDIES OF GENIUS

Edited by Lewis M. Terman

Volume I. Mental and Physical Traits of a Thousand Gifted Children

By Lewis M. Terman *and* Others

Volume II. The Early Mental Traits of Three Hundred Geniuses

By Catharine M. Cox

Volume III. The Promise of Youth: Follow-up Studies of a Thousand Gifted Children

By Barbara S. Burks, Dortha W. Jensen, *and* Lewis M. Terman

Volume IV. The Gifted Child Grows Up: Twenty-five Years' Follow-up of a Superior Group

By Lewis M. Terman *and* Melita H. Oden

Volume V. The Gifted Group at Mid-Life: Thirty-five Years' Follow-up of the Superior Child

By Lewis M. Terman *and* Melita H. Oden

Lewis Madison Terman
(1877-1956)

THE GIFTED GROUP AT MID-LIFE

THIRTY-FIVE YEARS' FOLLOW-UP
OF THE SUPERIOR CHILD

VOLUME V

GENETIC STUDIES OF GENIUS

BY

LEWIS M. TERMAN AND MELITA H. ODEN

STANFORD UNIVERSITY PRESS
STANFORD, CALIFORNIA
LONDON: OXFORD UNIVERSITY PRESS

1959

STANFORD UNIVERSITY PRESS
STANFORD, CALIFORNIA
LONDON: OXFORD UNIVERSITY PRESS

© 1959 BY THE BOARD OF TRUSTEES OF THE
LELAND STANFORD JUNIOR UNIVERSITY

PRINTED IN THE UNITED STATES OF AMERICA

LIBRARY OF CONGRESS CATALOG CARD NUMBER: 25-8797 REV.

ACKNOWLEDGMENT OF FINANCIAL ASSISTANCE

The research studies reported in this volume have been financed by grants-in-aid and anonymous gifts totaling close to $250,000.

The greater part of the expenses incurred between 1921 and 1929 were defrayed by the Commonwealth Fund of New York City. Stanford University, through the Thomas Welton Stanford Fund, financed the follow-up study of 1936–37 and contributed minor supplementary funds as needed between 1928 and 1936.

The Carnegie Corporation of New York provided two grants which made possible the extensive follow-up of 1939–41 and the statistical work on the resulting data between 1941 and 1943. The National Research Council, through its Committee for Research on Problems of Sex, financed the studies on marital adjustments reported in Volume IV, and the Columbia Foundation of San Francisco provided three annual grants which met approximately three-fourths of the expenses incurred between 1943 and 1946 in continued follow-up of the subjects and analysis of data. Annual grants from the Marsden Foundation of Palm Springs from 1946 to 1953 provided partial support of the study during those years.

Three grants from the Carnegie Corporation from 1949 to 1951 and a grant from the Rockefeller Corporation in 1950 financed the 1950–52 follow-up investigation. A research contract from the Office of Naval Research in 1951 defrayed the expenses of a study of scientists and non-scientists in the gifted group.

In 1955 a grant from the Fund for the Advancement of Education made possible the mail follow-up of that year and the preparation of much of the data for this volume.

Material assistance has been provided from time to time by gifts from various individual donors including several parents of the gifted subjects, a few of the subjects themselves, the owner of a well-known magazine, and other anonymous donors.

To all these benefactors, grateful acknowledgment of their support is made.

By special arrangement, all royalties from the publications resulting from this study have been assigned to the research funds.

FOREWORD

This report on the further development of the Terman Gifted Group brings them into mid-life. It provides a concise summary of the main findings reported in previous volumes, and then presents the data obtained in the field and mail follow-up studies that were completed in 1955. It tells, dramatically but objectively, what thirty-five years of living have done to those gifted eleven-year-olds whose characteristics were first described in Volume I of this series. If the interim evidence examined here leads to any one salient conclusion, it is that the end of their intellectual and social vigor is not yet in sight. Indeed, if ever there was excitement in figures, it is in those that form the data for this skillfully presented and gracefully written report of what these men and women look like in their maturity. For many a student of contemporary society, this volume will be the pay-off, the best and most informative of the series of *Genetic Studies of Genius*.

Professor Terman began this study in 1921, when he was in his mid-forties, and most of the children were about eleven. When he died near the end of his eightieth year, the "children" had reached their mid-forties themselves, and Professor Terman had maintained continuous data collection from them for more than three decades. In this time, he conducted three major field studies, several mail follow-ups, tested a large proportion of the group's offspring, and maintained such close personal relationships with the nearly 1,500 members of the group that 95 percent of them were still active research subjects.

The present volume was fully planned and the data analysis completed before Professor Terman's death. He had written initial drafts of Chapters I, II, III, and IX, and had made notes for Chapter XI; Mrs. Oden wrote the others, and prepared the final manuscript for publication. Thus ends a personal collaboration of many years and many volumes. It was a fortunate one, and one for which the collaborators' colleagues render much thanks.

What does not end is the lives and careers—and our study—of the gifted group. With data already in hand, Mrs. Oden will continue the analysis of developmental trends, including a number of investigations that had been projected by Professor Terman. The body of information provided by the gifted group to date is extremely valuable, not only for the findings already reported, but for analyses yet to be made. The future promises information of equal importance. Professor Terman foresaw this clearly, and before his death he completed arrangements with Stanford University for the continuing maintenance of the integrity and confidentiality of the files. He designated his son, Dr. Frederick E. Terman, now Provost of the University, and his long-time colleague, Dr. Quinn McNemar, Professor of Psychology at Stanford, as the official custodians. To me, he delegated the administrative responsibility for planning such continuing research operations with the group as may seem appropriate. To assist in this enterprise, he assigned funds from his estate that will provide partial support for maintaining the files for several years.

In this connection, a word is in order. Professor Terman began the project with a grant from the Commonwealth Fund in 1921. Since then, half a dozen other foundations and government agencies have contributed financial support. The total expended to date has been approximately a quarter million dollars. The largest single donor was Professor Terman himself, who provided more than one-fifth of the total cost by direct gift. In addition, he and his co-authors assigned to the study all royalties from their publications relating to the study.

And so we reach the end of one stage in this extraordinary research enterprise. When Professor Terman came to Stanford in 1910, as an assistant professor of education, the scientific study of the intellect had scarcely begun. In Paris, Alfred Binet had constructed an ingenious test for measuring academic ability in school children; at Columbia University's Teachers College, E. L. Thorndike had begun work on the measurement of school achievement. But it remained for Lewis Terman to conceive the development of a rigorous intelligence test that could select the ablest children and thus allow society to focus its full educative power on developing their potential. In 1916, Terman published the Stanford Revision of the Binet-Simon test. In the last years of World War I, he was a prime mover in the construction of the Army Alpha and Beta tests. After the war, with both individual and group tests available, he turned to the problem that had enthralled him ever since his graduate student days at Clark University. Although he made two useful excursions into other fields—the measurement of masculinity-femininity and of marital happiness—his main concern for the rest of his life was the research on gifted children.

What he started, now remains to be finished. Science is cumulative by its very nature, but only among the chroniclers of the stars and the waters have such prolonged studies of individual objects been made heretofore. We can be grateful for the courage and vision of the man who finally broke the barrier of the limited lifetime alloted to any one researcher, and got under way a study of man that will encompass the span of the *subjects'* lives, not just those of the researchers. So Professor Terman has bowed out, and this is the last interim report in which he will have participated personally. Certainly there will be other replacements in the list of our research personnel before the final volume of this series can be written. On actuarial grounds, there is considerable likelihood that the last of Terman's Gifted Children will not have yielded his last report to the files before the year 2010!

ROBERT R. SEARS

Stanford University
  January 10, 1959

PREFACE

This is the fifth volume in the *Genetic Studies of Genius* series and the fourth concerned with the longitudinal study of gifted children initiated by Lewis M. Terman in 1921.* As this report is being completed in the summer of 1958, more than 35 years have elapsed since the investigation was undertaken, and more than one and one-half years have passed since the death of Dr. Terman.

His imagination and vision, his fortitude and perseverance, his persuasiveness and charm—all these enabled him to follow the careers of this same group of subjects continuously for three and one-half decades until his death. It was due, in large measure, to the human qualities of Dr. Terman that his study of gifted children grew into a very close personal relationship between him and the members of his group—a relationship that has been maintained throughout the years. Always active in the research, he was able, after his retirement as head of the Psychology Department in 1942, to give it his full attention. His affection for the group is shown in the warm and personal tone of the follow-up letters, even the "form" letters, which bore the salutation *To my gifted "children,"* the quotation marks added in recognition of their adult status. It happens that the gifted children turned out, in one way or another, to be gifted adults, but if they had not, Dr. Terman's affection for them would have been no less. He repeatedly expressed, and not infrequently was called on to prove, his deep personal interest in each subject. As one of his letters accompanying a questionnaire put it, "although the published reports will be largely statistical, I want you to know that each of you is to me a real person and not just another statistic."

This study is unique in a number of ways: the many years it has continued, the vast amount of immensely valuable data collected, the zeal with which the inviolability of the records has been safeguarded and the confidences of the subjects respected, the enduring friendship and loyalty of the group, and their unparalleled cooperation. The fact that 95 percent of the group are still actively participating in the study, a striking manifestation of the rapport between investigator and sub-

---

* Vol. II of the series by Catharine M. Cox was titled *Early Mental Traits of Three Hundred Geniuses*[8] and is a collateral study dealing with historical geniuses.

jects, strengthens the validity of the findings. The debt of gratitude owed the subjects and their parents can never be adequately expressed.

The research has yielded documentary evidence that the gifted child is far more apt to become the intellectually superior, vocationally successful, well-adjusted adult than will the average. These findings have tremendous importance for education and it is a source of satisfaction to know that this study has contributed to the present interest in better educational provisions for gifted children everywhere. Dr. Terman's work stands as a landmark in the identification of superior mental ability and the factors that make for its effective utilization.

Another phenomenon of the study has been the constancy and devotion of those who have collaborated as assistants and consultants—testimony to Dr. Terman's gift for friendship, his generosity of spirit, loyalty and understanding. He has paid tribute in the previous volumes of this series to those who assisted him in the various phases of the study. The field study of 1950–52 was carried out by Nancy Bayley, Helen Marshall, Alice Leahy Shea, Ellen Sullivan, and myself. Dr. Marshall, who with Dr. Florence Goodenough conducted the original search for subjects in 1921–22, has the distinction of having worked in each follow-up investigation. Dr. Bayley and Dr. Sullivan had both assisted in the follow-up of 1939–40 and Dr. Shea had assisted in the field study of 1927–28. In an investigation in which success depends so much on personal relationships, the role of the field worker is an important one, and the study was fortunate that such experienced and highly qualified persons as Dr. Bayley, Dr. Marshall, Dr. Sullivan, and Dr. Shea were willing to take leaves of absence from their regular positions to assist in these follow-up studies.

Although I did not have the privilege of working on the initial investigation of 1921–22, I had known Dr. Terman even before that date in my student days. When he engaged me as one of his assistants in 1927 to help with the first field follow-up of the gifted group, it was the beginning of a long and close association. It is difficult to express or interpret in words the many intangible effects of this association. The inspiration, the intellectual stimulation, the new frontiers explored, the wise counsel, the warm friendship—all these were among the rewards of working with such a man.

Among those to whom the more recent phases of the study owe much are Dr. Olga W. McNemar, who assisted in the data analysis and preparation of reports between 1943 and 1949 and who was the chief collaborator with Dr. Terman in a study of scientists and nonscientists in the gifted group which was made in 1951–52. Dr. Quinn Mc-Nemar, who has been close to all of Dr. Terman's researches, has served as consultant on the gifted study for many years and has contributed constructive help on statistical problems growing out of the research.

Dr. Marian Ballin gave important statistical help in preparing the data for this volume.

Mrs. Babette Doyle, between the years 1947 and 1956, and Mrs. Shiela Buckholtz, from 1952 to the present, have each rendered invaluable assistance in the dual capacity of secretary and research assistant. In addition I am personally indebted to each for her help in the preparation of this manuscript through careful reading, criticism, and suggestions for improvement.

Finally, I am profoundly grateful to Dr. Robert Sears not only for his critical reading of the manuscript but also for his encouragement and faith when, with the loss of Dr. Terman, the responsibility for the completion of this volume fell to me. I have tried to make it as nearly as possible the book that Dr. Terman himself would have written. Lacking his grace and ease of expression as well as his wisdom and experience, I could not hope to succeed entirely. To whatever extent the goal has been achieved, special credit is due Mrs. Buckholtz, Mrs. Doyle, and Dr. Sears.

MELITA H. ODEN

TABLE OF CONTENTS

CHAPTER   PAGE

    I. EARLY HISTORY OF THE TERMAN STUDY OF GIFTED
       CHILDREN 1

   II. THE INTERMEDIATE YEARS 17

  III. THE GROUP REACHES MID-LIFE 23

  IV. MORTALITY, HEALTH, AND GENERAL ADJUSTMENT . 28

   V. INTELLECTUAL STATUS AT MID-LIFE 52

  VI. THE MATTER OF SCHOOLING 64

 VII. THE MATTER OF CAREER 73

VIII. AVOCATIONAL AND OTHER INTERESTS 107

  IX. SOME POLITICAL AND SOCIAL ATTITUDES 119

   X. MARRIAGE, DIVORCE, AND OFFSPRING 132

  XI. THE FULFILLMENT OF PROMISE 143

REFERENCES CITED 153

APPENDIX 157

INDEX 181

CHAPTER I

EARLY HISTORY OF THE TERMAN
STUDY OF GIFTED CHILDREN

Many philosophers and scientists from Plato and Aristotle to the present day have recognized that a nation's resources of superior talent are the most precious it can have. A number of factors, however, have operated to postpone until recent years the inauguration of research in this field. Among these are: (1) the influence of long-current beliefs regarding the essential nature of the genius, long regarded as qualitatively set off from the rest of mankind and not to be explained by the natural laws of human behavior; (2) the widespread superstition that intellectual precocity is pathological; and (3) the growth of pseudo-democratic sentiments that have tended to encourage attitudes unfavorable to a just appreciation of individual differences in human endowment.

The senior author's first exploration into the problems posed by intellectual differences occurred over a half-century ago when, as a graduate student, he made an experimental study of two small contrasting groups of bright and dull children.[34]* His interest was heightened a few years later when, in standardizing the 1916 Stanford-Binet Intelligence Scale, he located and studied about a hundred children whose IQ's were above 130. He then decided to launch, at the first opportunity, a large-scale investigation of the physical, mental, and personality traits of a large group of exceptionally gifted children and, by follow-up studies, to find out what kind of adults such children tend to become. It was obvious that no intelligent program for training the gifted child could be laid down until the answers to these questions had been found.

In 1921 a generous grant from the Commonwealth Fund of New York City made possible the realization of this ambition. The project as outlined called for the sifting of a school population of a quarter-

* Numerals refer to numbered and alphabetically arranged references on pages 153 ff.

million in order to locate a thousand or more of highest IQ. The subjects thus selected were to be given a variety of psychological, physical, and scholastic tests and were then to be followed as far as possible into adult life. The investigation was expected to tell us (1) what intellectually superior children are like as children; (2) how well they turn out; and (3) what are some of the factors that influence their later achievement.

The Selection of Subjects

The problem was to discover in the schools of California a thousand or more subjects with IQ's that would place them well within the highest one percent of the school population. For financial reasons it was not possible to give mental tests to the entire school population. Instead, the search was limited chiefly to the larger and medium-sized urban areas. The following procedures were used to identify the children of highest IQ in the areas surveyed.

In grades three to eight each classroom teacher filled out a blank which called for the name of the brightest child in the room, the second brightest, the third brightest, and the youngest. The children thus nominated in a particular school building were then brought together and given a group intelligence test (National Intelligence Test, Scale B). Those who scored promisingly high on the group test were given an individual examination on the Stanford-Binet test. In grades below the third, only the Stanford-Binet test was given to those nominated by the teacher, since no suitable group test was available at that time for younger children. In high schools the selection of subjects was based on the Terman Group Test scores of students nominated by the teachers as being among the brightest in their respective classes.

Checks made on the method of selection indicated that the method used was identifying close to 90 percent of all who could have qualified. The proportion was high enough to insure that the group selected for study constituted a reasonably unbiased sampling and that whatever traits were typical of these children would be reasonably typical of gifted children in any comparable school population. The original criterion for inclusion for the Binet-tested subjects was an IQ of 140 or above, but for various reasons sixty-five subjects were included in the IQ range of 135 to 139. Most of those below 140 IQ were either siblings of subjects already admitted to the group or were older subjects whose scores were deemed to be spuriously low because of insufficient top in the 1916 Stanford-Binet. The standard set was purely arbitrary and was intended to insure that the subjects included for study should be in the highest 1 percent of the school population in general intelligence as measured by the test used. Its choice was not based on any assumption that children above this IQ level are potential geniuses. The standards for admission on the Terman Group Test and other group tests also required the subject to score within the top 1 percent of the general school population on which the norms were established.

The nature and results of the early stages of the investigation have been fully described in an earlier publication[39] and will be summarized in the following pages.

Composition of the Group

The gifted subjects whose careers we have followed number, in all, 1,528 (857 males and 671 females). This figure includes a few who were selected before 1921, and 58 who were not selected until the field study of 1927–28. These 58 were siblings of previously selected subjects who were too young to test at the time of the main search for subjects in 1921–22.

The Binet-tested group made up more than two-thirds of the total and included 1,070 subjects (577 boys and 493 girls). Selected by the Terman Group Test given in high schools were 428 subjects (265 boys and 163 girls). The remaining 30 subjects were chosen on the basis of scores on the National Intelligence Test or the Army Alpha Test.* The average age of the total group at the time of selection was 11 years; the Binet-tested subjects averaged 9.7 years and those qualifying on a group test, 15.2 years.

The mean IQ of subjects who were given the Stanford-Binet was 151.5 for the boys, 150.4 for the girls, and 151.0 for the sexes combined. The IQ range was from 135 to 200 with 77 subjects scoring at IQ 170 or higher. The mean IQ of high-school subjects tested by the Terman Group Test was 142.6 and the range of IQ was from 135 to 169. These figures, however, were estimates based upon norms which were inadequate and were perhaps 8 or 10 IQ points too low. Later follow-up of the high-school subjects indicated that they were as highly selected as the Binet-tested group.

* This group includes 24 pre-high-school pupils with National Intelligence Test scores and 6 high-school students with Army Alpha Test scores who were not tested in the formal search for subjects, but were brought to the attention of the study by their schools and included because of their very high test scores.

The sex ratio among the Binet-tested subjects was approximately 116 boys to 100 girls. The much higher sex ratio for the high-school subjects—roughly 160 boys to 100 girls—is probably due to the less systematic procedures used in locating gifted subjects in the high schools. A sex ratio of 116 males to 100 females may be fully accounted for by the greater variability of males. McNemar and Terman,[27] in a survey of sex differences on variability in such tests as the Stanford-Binet, the National Intelligence Tests, the Pressey Group Test, and Thorndike's CAVD test, found that 29 of 33 sex comparisons based on age groupings showed greater variability of boys. In Scotland, 874 of 875 children who were born on four particular days of the calendar year 1926, and were still living in 1936, were given a Stanford-Binet test at the age of ten years. The S.D. of the IQ distribution for this perfect sample was 15.9 for boys and 15.2 for girls—a difference sufficient to give a sex ratio of 134 boys to 100 girls scoring as high as 140 IQ.

KINDS OF INFORMATION OBTAINED

Besides the intelligence test scores on which the selection of subjects was based, information of many different kinds was obtained. The chief sources were as follows.

1. A twelve-page Home Information Blank was filled out by the child's parents. This called for information on developmental case history, circumstances of birth, early feeding, ages of walking and talking, illnesses, nervous symptoms, home training, indications of intelligence, age of learning to read, reading habits, educational and occupational achievement of parents, genealogical records, and ratings on twenty-five traits.

2. An eight-page School Information Blank was filled out by the child's teacher. The blank called for information on school health records, quality of school work in each separate subject, evidence of superior ability, amount and kinds of reading, nervous symptoms, social adjustment, and ratings on the same twenty-five traits that were rated by the parents. This information was also obtained for a control group of 527 unselected school children.

3. A one-hour medical examination was given to 783 gifted subjects. The examination covered vision, hearing, nutrition, posture, teeth, heart, lungs, genitals, glandular disorders, blood pressure and

hemoglobin tests, pulse and respiration rates, urine tests, and neuro-logical conditions.

4. Thirty-seven anthropometrical measurements were made of nearly 600 gifted subjects.

5. A three-hour battery of achievement tests was given to 550 gifted subjects in grades two to eight. The battery covered reading, arith-metical computation, arithmetical reasoning, language usage, spelling, science information, language and literature information, history and civics information, and art information. The same tests were given to a large control group of unselected subjects.

6. A four-page Interest Blank was filled out by all the gifted sub-jects who were able to read and write and by a large control group of unselected subjects. The blank called for information on occupational preferences, reading interests, school-subject interests, relative diffi-culty of school subjects, number and size of collections, and various activities and accomplishments.

7. A record of all books read over a period of two months was obtained from some 550 gifted subjects and from a control group of 808 unselected children. Each book read was rated by the child for degree of interest.

8. A test of play interest, play practice, and play information was given to all the gifted subjects above grade two, and to a control group of nearly 500 unselected children. This test yielded scores on mascu-linity, maturity, and sociability of interests, and a play information quotient.

9. A battery of seven character tests was given to 550 gifted sub-jects and 533 unselected children of a control group. These included two tests of overstatement; three tests of questionable interests, prefer-ences, and attitudes; a test of trustworthiness under temptation to cheat; and a test of emotional stability.

FAMILY BACKGROUND

All racial elements in the areas covered were represented in the group, including Orientals, Mexicans, and Negroes. They came from all kinds of homes, from the poorest to the best, but the majority were the offspring of intellectually superior parents. The tendency to supe-riority in the social and cultural background of the subjects is shown in many ways. Nearly a third of the fathers as of 1922 were in professional occupations, and less than 7 percent in semiskilled or unskilled work. The mean amount of schooling of both fathers and mothers was approximately 12 grades, or about four grades more than the average person of their generation in the United States. A third of the fathers and 15.5 percent of the mothers had graduated from college. Twenty-eight fathers and six mothers had taken a Ph.D. degree—numbers which were considerably increased later. By 1940 the number of parents listed in *Who's Who in America* were 44 fathers and 3 mothers.

The number of books in the parents' homes, as estimated by the field assistants, ranged from almost none to 6,000, with one home out of six having 500 or more. The median family income during 1921 for a random sample of 170 families in the group was $3,333; the average for the sample was $4,705. Only 4.4 percent reported $1,500 or less, while 14.1 percent reported $8,500 or more, and 4.1 percent reported $12,500 or more. The field assistants rated a random sample of 574 homes on the Whittier Scale for Grading Home Conditions. Rating superior to very superior were 60.3 percent, as contrasted with 9.5 percent rating inferior to very inferior.

Additional evidence of the superiority of family background is the fact that 182 of the families contributed two or more subjects to the group. Among these were 2 families of five children, all of whom qualified for the gifted group, 10 families each of whom contributed four children to the group, and 20 families who contributed three children each to the group. There were also 28 families whose children, often two or more in a family, are first cousins. Since not more than one child in a hundred of the general school population could qualify for the group, the likelihood that two such children would be found in one family would be almost infinitesimal by the laws of pure chance. That so many families contributed two or more children to the group means that something besides chance was operating, such as common ancestry, common environment, or, more probably, both of these influences.

Physique and Health of Gifted Children

Anthropometric measurements were made of a random gifted group of 312 boys and 282 girls, all but a few of whom were between the ages of 7 and 14. The results showed that the gifted children as a group exceeded the best standards at that time for American-born children in growth status as indicated by both height and weight, and that they were also above the established norms for unselected children in California.

Information on physical history was obtained from parents and teachers for nearly all the subjects, and information on health history was also obtained from the teachers for a control group of 527 unselected children enrolled in the classes attended by members of the gifted group. The mean birth weight reported by the mothers of the gifted was about three-quarters of a pound above the norm according to the commonly accepted standards of the time. About 17 percent of male births and 12 percent of female births involved instrumental deliveries; these rather high figures probably reflect the quality of obstetrical service obtained by parents of superior intelligence and above-average income. The proportion of breast feeding was considerably in excess of the figures reported by Woodbury[44] for the general population. The reported ages of learning to walk averaged about a month less, and the age of learning to talk about three and one-half months less, than the mean ages reported for unselected children. Among the older children, the onset of puberty, as indicated by change of voice in boys and by first menstruation of girls was, on the average, earlier than for children of the general population. About a third of the gifted subjects had suffered one or more accidents, 8 percent having had bone fractures. The number of surgical operations averaged one per child, over half of which were for adenoids or tonsils.

The School Information Blank filled out by teachers of the gifted subjects, and also for a control group attending the same classes, furnished interesting comparative data. These reports indicate that "frequent headaches" were only half as common among the gifted as among the controls, "poor nutrition," a third as common, "marked" or "extreme" mouth-breathing two-thirds as common, and defective hearing half as common. The two groups did not differ significantly with respect to frequency of colds, "excessive timidity," or "tendency to worry," but "nervousness" was reported for 20 percent fewer gifted than controls.

The medical examination was given to 783 of the gifted subjects who lived in or near Los Angeles or the San Francisco Bay area. The examinations were made by two experienced child specialists, both of whom had had two years of postgraduate work in the department of pediatrics at the University of California Medical School. All examinations were made in the physician's office, to which the child was brought by a parent, usually the mother. The incidence of physical defects and abnormal conditions of almost every kind was below that usually reported by school physicians in the best medical surveys of school populations in the United States. This is certainly true for defects of hearing and vision, obstructed breathing, dental caries, malnutrition, postural defects, abnormal conditions of the heart or kidneys, enlargement of the bronchial glands, and tuberculosis. The sleep and dietary regimes of the group as a whole were found to be definitely superior. The incidence of nervous habits, tics, and stuttering was about the same as for the generality of children of corresponding age. The examining physicians, notwithstanding occasional disagreement in their results, were in complete accord in the belief that, on the whole, the gifted children of this group were physically superior to unselected children.

The combined results of the medical examinations and the physical measurements provide a striking contrast to the popular stereotype of the child prodigy so commonly depicted as a pathetic creature, overserious and undersized, sickly, hollow-chested, stoop-shouldered, clumsy, nervously tense, and bespectacled. There are gifted children who bear some resemblance to this stereotype, but the truth is that almost every element in the picture, except the last, is less characteristic of the gifted child than of the mentally average.

EDUCATIONAL HISTORY

The average age on entering school (above kindergarten) was six and a quarter years. Low first grade was skipped by 21 percent of the children, and the entire first grade by 10 percent. The average progress quotient for the entire gifted group was 114, which means that the average child was accelerated to the extent of 14 percent of his age. According to the testimony of their teachers, the average gifted child merited additional promotion beyond where he was by 1.3 half-grades. "Strong" liking for school was reported by parents for 54 percent of boys and 70 percent of girls, as compared to only 5 percent of the sexes combined for whom parents reported either "slight liking" or "positive dislike."

Nearly half of the children learned to read before starting to school; 20 percent did so before the age of five years and 6 percent before four years. Other early indications of superior intelligence most often mentioned by parents were quick understanding, insatiable curiosity, extensive information, retentive memory, large vocabulary, and unusual interest in number relations, atlases, and encyclopedias.

The Stanford Achievement Tests were given in the spring of 1922 to a random group of 565 gifted children in grades below the ninth. The tests provided separate scores for reading, computation, arithmetical reasoning, language usage, spelling, and four different fields of information. The average achievement quotient for the school subjects combined was 144, and only one quotient in six was as low as 130. The difference of 30 quotient points between the average achievement quotient of 144 and the average progress quotient of 114 (noted above), means that the average gifted child was retarded in grade placement by 30 percent of his age below the level of achievement which he had already reached. More than half of those tested had mastered the school curriculum to a point two full grades beyond the one in which they were enrolled, and some of them as much as three or four grades beyond. For the fields of subject matter covered by our tests, the superiority of the gifted subjects over unselected children was greatest in reading, arithmetical reasoning, and information; it was least in computation and spelling.

Another question answered by the achievement tests was whether the gifted child tends to be more one-sided in his abilities than the average child, as so many people believe to be the case. Analysis of the subject-matter achievement quotients of the gifted group as compared to a group of unselected children shows that the amount of unevenness in subject-matter profiles of the gifted does not differ significantly from that shown by unselected children.

CHILDHOOD INTERESTS AND PREOCCUPATIONS

The four-page Interest Blank was filled out by all of the gifted who were old enough to read, and also by a control group of unselected children. In a long list of school subjects the children were asked to rate on a five-point scale their liking for each of the school subjects they had studied. We will consider here only the ratings given by children of ages 11 to 13, inclusive. Analysis of the ratings showed that gifted children were more interested than were unselected children in school subjects which are most abstract, and somewhat less interested in the "practical" subjects. Literature, debating, dramatics, and history were rated much more interesting by the gifted, while penmanship, manual training, drawing, and painting were rated somewhat higher by the control group. When cross-sex comparisons were made, it was found that in their scholastic interests gifted girls resembled gifted boys far more closely than they resembled control girls.

In the same Interest Blank was a list of 125 occupations and the child was told to place one cross before each occupation he might possibly wish to follow and two crosses before his one first choice. The data were treated for ages 8 to 13 for both gifted and control groups. Analysis of the data revealed that the gifted showed greater preference for professional and semiprofessional occupations, and the control group greater preference for mechanical and clerical occupations and for athletics.

The test of interest in 90 plays, games, and other activities was designed in such a way as to yield a preference score on each of the 90 items for each age and sex group of gifted and control subjects. Comparison of boys and girls in the control group with respect to kinds of plays and games preferred made it possible to derive a masculinity-femininity index for each child. Similarly, by comparison of preferences expressed at different ages in the control group, an index of interest maturity was derived for each child. Finally, an index of sociability was computed which was based on the extent of a child's preference for plays and games that involve social participation and social organization. Comparison of the gifted and control (i.e., unselected) children on these three indices yielded the following conclusions: (1) Gifted boys tended to be somewhat more masculine in their play interests than control boys at all ages from 8 to 12 years, after which there was little difference. Gifted and control girls did not differ significantly at ages 8, 9, and 10, but at ages 11, 12, and 13 the gifted girls tended to be more masculine. (2) Comparisons of maturity indices for gifted and control subjects showed greater maturity of play interests for the gifted of both sexes at all age levels; i.e., they were ahead of their years in play interests. (3) Comparisons on sociability indices showed gifted subjects of both sexes significantly below control subjects at all ages; i.e., age for age the control subjects had somewhat more interest than gifted subjects in plays that involve social participation. Much of this difference can be accounted for by the fact that the gifted child is more self-sufficient and thus more able to amuse himself.

A test of play information (composed of 123 items that could be scored objectively) was devised which yielded a play information quotient based on age norms for unselected children. The average play information quotient of the gifted was 137, and only 3 percent of the group were below 100. The average gifted child of 9 years had acquired more factual information about plays and games than the average un-selected child of 12 years.

Information on the amount and kind of reading done was obtained by having 511 gifted children and 808 children of a control group keep a record of each book read during a period of two months. The records revealed that the average gifted child was reading about 10 books in two months by age 7, and 15 books by age 11, with little increase there-after. Few of the control group read any books below 8 years, and after 8 years the average number read in two months was less than half that of the gifted. Classification of the books read showed the gifted children reading over a considerably wider range than the control children. The gifted, much more often than the control group, preferred science, his-tory, biography, travel, poetry, drama, and informational fiction.

CHARACTER TESTS

Do children of superior intelligence tend to be superior also in char-acter traits? An answer to this question was sought by giving a battery of seven character tests to a random group of 532 gifted children aged 7 to 14 years and to a control group of 533 unselected children aged 10 to 14 years. The battery included two tests of the tendency to overstate in reporting experience and knowledge; three tests of the wholesome-ness of preferences and attitudes (reading preferences, character pref-erences, and social attitudes, respectively); a test of cheating under circumstances that offered considerable temptation; and a test of emo-tional stability. The tests were so devised that they could be scored objectively and could be given to the subjects in groups. The nature of the several tests is described in an earlier report.[39]

The tests of cheating and emotional stability were selected as among the best of a battery of character tests used by Cady;[4] the others were all from a battery devised by Raubenheimer.[29] Both of these batteries had been found to yield satisfactory reliability coefficients and to dis-criminate rather effectively between boys of known delinquent tenden-cies and boys of superior social and behavioral adjustment. Total scores of the seven character tests have a reliability of .80 to .85 and a validity (based on discrimination between delinquent and well-adjusted boys of ages 12 to 14) of approximately .60. Whether the validity is equally high for girls is not known.

The results of the character tests were decisive; the gifted group scored "better" than the control group on every subtest at every age from ten to fourteen. Below the age of ten no comparison was possible because the control subjects below this age were not sufficiently literate to take the tests. Table 1 shows, for the sexes separately, the proportion of gifted subjects who equaled or surpassed the mean of the control group on each subtest and on the total score for ages ten to fourteen combined.

TABLE 1

PROPORTION OF GIFTED SUBJECTS WHO EQUALED OR SURPASSED
THE MEAN OF CONTROL SUBJECTS IN EACH OF SEVEN
CHARACTER TESTS AND IN TOTAL SCORE

Tests	Boys %	Girls %
1. Overstatement A	57	59
2. Overstatement B	63	73
3. Book preferences	74	76
4. Character preferences	77	81
5. Social attitudes	86	83
6. Cheating tests	68	61
7. Emotional stability	67	75
Total score	86	84

The question may be raised whether a part of the superiority is spurious because of the possibility that bright subjects would be more likely to divine the purpose of the tests and so respond in the socially approved way. This factor, if present at all, would be most likely to influence scores on reading preferences, character preferences, social attitudes, and emotional instability. It is believed hardly to have entered at all in the cheating test (disguised as a test of motor accuracy) or in the two overstatement tests, all three of which gave highly reliable differences between the gifted and control groups. In his study of delinquent and well-adjusted boys, Raubenheimer[29] questioned his subjects after they had completed the tests, to find out whether they had guessed their purpose. Less than 5 percent of his subjects (all thirteen years old) guessed correctly.

TRAIT RATINGS

The plan of trait rating used with the gifted subjects was the result of several years' experience in trying out various rating schemes with children of average and superior ability. The traits finally selected for rating numbered 25 and can be classified in the following categories:

intellectual (4), volitional (4), moral (4), emotional (3), aesthetic (2), physical (2), social (5), and the single trait, mechanical ability. The individual traits are listed by category in Table 2. However, in the blanks in which the ratings were made, the traits were presented in a mixed order.

A cross-on-line technique was used in getting the rating for each trait and the ratings were scored in intervals of 1 to 13. Nearly all of the gifted subjects were rated both by a parent and by a teacher. Teacher ratings were also obtained for 523 children of ages 8 to 14 in a control group composed of unselected children enrolled in the same classes as the gifted.

Parents and teachers agreed fairly well regarding the traits on which the gifted children were most or least superior to average children. The rank order of the traits from highest to lowest mean rating by parents correlated .70 with the corresponding rank order based on teachers' ratings. However, the agreement was much less in their ratings on individual children; for most of the traits it was represented by a Pearsonian correlation of only about .30. This figure should not be regarded as a reliability coefficient in the true sense, for the reason that a child's personality behavior in the school is often very different from that which he exhibits in the home.

More important is the comparison of gifted and control subjects on ratings given to both groups by the teachers. Table 2 gives the comparative data on both for the 25 individual traits and for groups of traits as classified in various categories. The figures in Table 2 are for the sexes combined and for all ages combined, since the mean ratings varied only slightly either with age or with sex. The slight variation by age and sex was to be expected in view of the fact that raters were instructed to rate each subject in comparison with the "average child of his age and sex."

The superiority of the gifted over the control subjects as shown by teachers' ratings agrees fairly well with the data from other sources. This is especially true in regard to the kinds of traits in which the superiority of the gifted is most or least marked. At the top of the list are the four intellectual traits, with 89 percent of gifted rated at or above the mean of control subjects. Especially high were the ratings of "general intelligence" and "desire to know." Next highest were the four volitional traits, with percentages in the narrow range of 84 to 81. Third highest are the three emotional traits, with "sense of humor" (74%) the

TABLE 2

PERCENTAGES OF GIFTED SUBJECTS RATED BY TEACHERS
ABOVE THE MEAN OF THE CONTROL GROUP

Percent

1. *Intellectual traits*

General intelligence	97	
Desire to know	90	
Originality	85	
Common sense	84	
Average of intellectual traits		89

2. *Volitional traits*

Will power and perseverance	84	
Desire to excel	84	
Self-confidence	81	
Prudence and forethought	81	
Average of volitional traits		82.5

3. *Emotional traits*

Sense of humor	74	
Cheerfulness and optimism	64	
Permanence of moods	63	
Average of emotional traits		67

4. *Aesthetic traits*

Musical appreciation	66	
Appreciation of beauty	64	
Average of aesthetic traits		65

5. *Moral traits*

Conscientiousness	72	
Truthfulness	71	
Sympathy and tenderness	58	
Generosity and unselfishness	55	
Average of moral traits		64

6. *Physical traits*

Health	60	
Physical energy	62	
Average of physical traits		61

7. *Social traits*

Leadership	70	
Sensitivity to approval	57	
Popularity	56	
Freedom from vanity	52	
Fondness for large groups	52	
Average for social traits		57.4

8. *Mechanical ingenuity* 47

highest of the three. The two aesthetic traits rank fourth with percentages of 64 and 66. Of the four moral traits, "conscientiousness" and "truthfulness" are rated reliably higher than the other two ("sympathy" and "generosity"). The two ratings on physical traits, which rank next, agree fairly well with the physical data obtained from medical examinations, health histories, and anthropometric measurements. Ranking seventh are the five social traits; of these, only "leadership" (with 70%) is rated very much above the mean of control children. The ratings on "leadership" are consistent with the later follow-up studies which have shown the high frequency with which gifted subjects have been elected to class offices and honors despite their usual age disadvantage.

"Mechanical ingenuity" was the one trait in which teachers rated the gifted below unselected children. It is certain that the teachers were in error here, for test scores in mechanical ability have been consistently found to yield positive, not negative, correlations with intelligence scores. This is a trait which the average classroom teacher has little opportunity to observe; moreover, she is prone to overlook the fact that the gifted child in her class is usually a year or two younger than the others.

Summary Portrait of the Typical Gifted Child

Although there are many exceptions to the rule, the typical gifted child is the product of superior parentage—superior not only in cultural and educational background, but apparently also in heredity. As a result of the combined influence of heredity and environment, such children are superior physically to the average child of the general population.

Educationally, the typical gifted child is accelerated in grade placement about 14 percent of his age; but in mastery of the subject matter taught, he is accelerated about 44 percent of his age. The net result is that during the elementary-school period a majority of gifted children are kept at school tasks two or three full grades below the level of achievement they have already reached.

The interests of gifted children are many-sided and spontaneous. The members of our group learned to read easily and read many more and also better books than the average child. At the same time, they engaged in a wide range of childhood activities and acquired far more knowledge of plays and games than the average child of their years. Their preferences among plays and games closely follow the normal sex trends with regard to masculinity and femininity of interests, although gifted girls tend to be somewhat more masculine in their play life than the average girls. Both sexes show a degree of interest maturity two or three years beyond the age norm.

A battery of seven character tests showed gifted children above average on every one. On the total score of the character tests the typical gifted child at age 9 tests as high as the average child at age 12.

Ratings on 25 traits by parents and teachers confirm the evidence from tests and case histories. The proportion of gifted subjects rated superior to unselected children of corresponding age averaged 89 percent for four intellectual traits, 82 percent for four volitional traits, 67 percent for three emotional traits, 65 percent for two aesthetic traits, 64 percent for four moral traits, 61 percent for two physical traits, and 57 percent for five social traits. Only on mechanical ingenuity were they rated as low as unselected children, and this verdict is contradicted by tests of mechanical aptitude.

Three facts stand out clearly in this composite portrait: (1) The deviation of gifted children from the generality is in the upward direction for nearly all traits; there is no law of compensation whereby the intellectual superiority of the gifted is offset by inferiorities along nonintellectual lines. (2) The amount of upward deviation of the gifted is not the same for all traits. (3) This unevenness of abilities is no greater for gifted than for average children, but it is different in direction; whereas the gifted are at their best in the "thought" subjects, average children are at their best in subjects that make the least demands upon the formation and manipulation of concepts.

Finally, the reader should bear in mind that there is a wide range of variability within our gifted group on every trait we have investigated. Descriptions of the gifted in terms of what is typical are useful as a basis for generalization, but emphasis on central tendencies should not blind us to the fact that gifted children, far from falling into a single pattern, represent an almost infinite variety of patterns.

CHAPTER II

THE INTERMEDIATE YEARS

For several years after 1922 the progress of the group was followed by means of information blanks sent annually to parents and teachers requesting certain physical, educational, and social data. In 1927–28 a more thorough investigation by field workers was undertaken and became the first of three field follow-ups at approximately 10- to 12-year intervals. These field studies were supplemented by intervening surveys by mail.

Six Years Later: The Promise of Youth

The six-year interval between the original research and the follow-up investigation of 1927–28 was in a number of respects favorable as to length; it was great enough to make a comparison between earlier and later findings significant and interesting, but not so long as to make it impossible to use any of the kinds of tests employed in the original study.

At the time of the follow-up (1927–28) the average age of the subjects was between 16 and 17 years and the majority were in high school. The data secured for the subjects included intelligence tests, school achievement tests, personality tests, and interest tests. Other types of data obtained were as follows: a Home Information Blank of four pages was filled out by parents of subjects up to and including age nineteen. A two-page Interest Blank was filled out by the subjects under twenty, and a two-page Information Blank by those twenty or over. A School Information Blank of two pages was filled out by the teachers of the children who were still in elementary or high school. A Trait Rating Blank provided ratings by parents and teachers on 12 traits selected from the 25 on which ratings were secured in 1921–22. Finally, blanks were provided for the field workers' reports on home visits and on conferences with the children themselves and their teachers. It was not possible, unfortunately, to repeat the medical examinations and physical measurements of the original study, but considerable information on physical development and health history was secured from parents and teachers.

Perhaps the most important outcome of the 1927–28 follow-up was the fact that the composite portrait of the group had changed only in minor respects in six years. As a whole, the group was still highly superior intellectually, for the most part within the top 1 or 2 percent of the generality. There was some evidence that the boys had dropped slightly in IQ and that the girls had dropped somewhat more. This conclusion, however, needs to be qualified in two respects: For one thing, the intelligence tests used in the follow-up lacked sufficient top to yield IQ's strictly comparable with those of 1921–22; for another, it should be pointed out that some regression toward the mean is to be expected from purely statistical considerations.

The showing in school achievement was in line with that for intelligence. There was less skipping of school grades after the age of eleven or twelve years, but the quality of work for the group in general remained at an exceedingly high level. For example, nearly two-thirds of the high-school grades of the girls and more than one-half of the high-school grades of the boys were A's. The significance of this is accentuated by the fact that the gifted group in the high-school period averaged considerably younger than the generality of high-school students. In evaluating school achievement at the high-school or college level, it is also necessary to bear in mind that the higher the grade the more highly selected is the school population with whom the gifted subjects are compared.

The composite-portrait method is useful, just as concepts and generalizations are useful in the shorthand of thinking. Nevertheless, the composite portrait, like any other kind of average, fails to convey any sense of the uniqueness of the individual subjects who compose the group. Although deviations below average intelligence were not found in the 1927–28 follow-up, extreme deviations both from the group average and from the generality were found in almost every physical, mental, and personality trait, including size, athletic ability, health, scientific ability, literary ability, masculinity, social and activity interests, vocational aptitude, social intelligence, leadership, ambition, and moral dependability. It is true that on all of these traits the mean for the gifted group tends to be higher than for unselected children of corresponding age, but the range of variability in these and other traits was if anything greater in mid-youth than it had been in mid-childhood. The results of the 1927–28 follow-up and a collateral study of the literary juvenilia of certain members of the group are described at length in Volume III of this series.[3]

EIGHTEEN YEARS LATER: THE GIFTED CHILD GROWS UP

For eight years after the follow-up described in the preceding section there was no systematic attempt to contact all the members of the gifted group. During that period, however, considerable correspondence was carried on with the parents and occasionally with the individual subjects of the group. Many wrote about their activities, or came to Stanford University for personal interviews. From the majority of the group, however, little information was secured during this period.

In 1936 plans were laid for an extensive field study to be made as soon as funds should become available. First, however, it seemed desirable to get in touch with as many as possible of the original group by mail, and to secure certain information that would aid in planning for the projected field study. Accordingly, a letter was sent out detailing the study and its purposes, and asking for the address of the subject, of his parents, and of a relative or friend through whom the subject might be located in later years. The letter was usually sent to the parents, although occasionally it went to the subject himself at the most recent address in our files. When the addresses had been received, a four-page Information Blank was sent to each subject, and a four-page Home Information Blank to the parents or, if both parents were deceased, to a near relative. The blank was accompanied by a letter emphasizing our continued interest in the subject, and the value both to science and to education of exact knowledge regarding the adult careers of persons who had tested high in intelligence during childhood.

The subject's Information Blank called for detailed information regarding educational history, occupations since leaving school, avocational interests and activities, general health, marital status, and deaths among relatives since 1928. The Home Information Blank called for information on the subject's physical and mental health, indications of special abilities, personality and character traits, education and occupations of siblings, and the accomplishments and activities of the subject's parents. Both blanks gave ample space for "additional information" not called for by specific questions and this brought, in many cases, extremely valuable and detailed replies.

The blanks were sent out in the spring of 1936 and, though the great majority of the reports had been received by midsummer, there was considerable difficulty in locating some of the subjects. It was nearly a year before blanks could be placed in the hands of all those who finally were located. This mail follow-up was more successful than had been expected, for approximately 90 percent of the subjects were located and from nearly all of them considerable information was obtained.

The information collected in 1936 brought the case history records up to date and thus set the stage for the more searching investigation of 1939–40. This follow-up was of special interest because it constituted the first person-to-person contact with the subjects since they had become adults (the average age in 1940 was 29.5 years). At the time of the original survey and again in the 1927–28 field follow-up, except for the tests and personal interviews which were conducted chiefly at the schools, most of our information had been obtained from the parents and the teachers. This was intentional because it was our purpose not to emphasize the study and its implications in the minds of the children. The following quotation from the instructions to the parents and teachers both in 1921–22 and in 1927–28 is self-explanatory:

In order to avoid the danger of causing undue self-consciousness, parents are urged not to call the child's attention to the fact that he (or she) is being studied. Do not tell the child the exact result of the mental test, or say anything about the special information which is sought in this blank. Publicity of every kind should be avoided. Do nothing which could possibly stimulate vanity or self-consciousness.

Although many—perhaps most—of the children did know by 1927–28 the nature of the experiment in which they were involved at least to the extent of realizing they had been chosen as bright students by a professor who was interested in such students, the knowledge seems not to have been of much concern to most. This was brought out in the Information Blank of 1936 which asked this question: *What effects (favorable, unfavorable, or both) has this knowledge (of being a subject in an investigation of gifted children) had upon you?*

The responses classified separately for the sexes were as follows:

	Men %	Women %
Favorable	12.9	18.7
Unfavorable	9.1	10.7
Both favorable and unfavorable	4.7	6.3
No effect	73.3	64.3

The second field follow-up began late in 1939 and continued into 1941 but the bulk of the data was collected in 1940. Four field workers spent the year interviewing the subjects and, wherever possible, their parents also. Intelligence tests were administered to the subjects, their spouses, and their offspring and extensive questionnaire data were collected from the subjects, the spouses, and the parents of the subjects. In addition, the Strong Vocational Interest Blank[31] was filled out by the men of the group. The study was highly successful with some information secured either directly or indirectly for nearly 98 percent of the living subjects, and fairly complete data were furnished by the 96 percent who co-operated actively.

So great was the amount of material on hand at the close of the field study that the punched card technique was used to analyze and correlate the data more efficiently. Not only the 1940 information but also the extensive case history data accumulated since the inception of this research were coded and transferred to punched cards. This method of handling the data made possible a very detailed study of the gifted child grown up, which has been fully reported in the preceding volume of this series.[36] This earlier volume includes also the results of a supplementary survey conducted by means of a questionnaire mailed in 1945–46, which brought up to date the records on the main events in the lives of the subjects between 1940 and 1946—close to 25 years after they had been selected for study.

In addition to the reports on mortality, general health and physique, mental health and general adjustment, intellectual status, educational histories, occupational status and income, vocational interest tests, avocational interests, political and social attitudes, marriage and offspring, and marital adjustment, Volume IV includes several special studies based on an analysis of total case history. The studies included subjects of IQ 170 and above, subjects of Jewish descent, factors in the achievement of gifted men, and the effects of school acceleration. Among the conclusions reached were the following:

That to near mid-life, such a group may be expected to show a normal or below-normal incidence of serious personality maladjustment, insanity, delinquency, alcoholism, and homosexuality.

That, as a rule, those who as children tested above 170 IQ were more often accelerated in school, got better grades, and received more schooling than lower-testing members of the group; that they are not appreciably more prone to serious maladjustment; and that vocationally they are more successful.

That gifted children who have been promoted more rapidly than is customary are as a group equal or superior to gifted nonaccelerates in health and general adjustment, do better school work, continue their education further, and are more successful in their later careers.

That the intellectual status of the average member of the group at the mean age of thirty years was close to the 98th or 99th percentile of the general adult population, and was far above the average level of ability of graduates from superior colleges and universities.

That in vocational achievement the gifted group rates well above the average of college graduates and, as compared with the general population, is represented in the higher professions by eight or nine times its proportionate share.

That the vocational success of subjects, all of whom as children tested in the top 1 percent of the child population is, as one would expect, greatly influenced by motivational factors and personality adjustment.

That the incidence of marriage in the group to 1945 is above that for the generality of college graduates of comparable age in the United States, and about equal to that in the general population.

That marital adjustment of the gifted, as measured by the marital happiness test, is equal or superior to that found in groups less highly selected for intelligence, and that the divorce rate is no higher than that of the generality of comparable age.

That the sexual adjustment of these subjects in marriage is in all respects as normal as that found in a less gifted and less educated group of 792 married couples.

That the test of marital aptitude predicts later marital success or failure in this group a little better than the test of marital happiness, much better than the index of sexual adjustment, and almost as well as scholastic aptitude tests predict success or failure in college.

That offspring of gifted subjects show almost exactly the same degree of filial regression as is predicated by Galton's Law.

That the fertility of the group to 1945 is probably below that necessary for the continuation of the stock from which the subjects come, and that this stock is greatly superior to the generality.

That Jewish subjects in the group differ very little from the non-Jewish in ability, character, and personality traits, as measured either by tests or by ratings, but that they display somewhat stronger drive to achieve, form more stable marriages, and are a little less conservative in their political and social attitudes.

CHAPTER III

THE GROUP REACHES MID-LIFE

The third field follow-up of the gifted subjects was made in 1950–52. In order to pave the way for the field worker contacts, a General Information Blank was mailed in the spring of 1950, and had been returned by the majority of the subjects by the fall of that year. The field work got under way in late 1950 and was completed about mid-1952.

The preliminary blank called for some twenty kinds of information that would furnish a profile of the gifted subjects at mid-life. Other blanks and tests used in gathering follow-up data included the following: a highly difficult test of intelligence (Concept Mastery test) comparable to that used in 1939–40 but with certain improvements; an abbreviated form of the 1939–40 marital happiness test; a four-page questionnaire calling for information on factors that had influenced rate of reproduction; an eight-page questionnaire designed to throw light on factors relating to childhood and family background that might have influenced personality development, motivation, or life success; a four-page blank relating to the development, health history, and personality characteristics of each child born to members of the group.

The field work program included personal interviews with as many of the subjects as possible (and usually the spouses also) who were living in California, the administration of the Concept Mastery test and the various supplementary questionnaires to the subject and the spouse, and the testing of offspring of appropriate age with the Stanford-Binet test. In the case of those subjects living at a distance (about 18% of the total), funds were not sufficient to provide for visits by field workers. Except for the personal interview and the intelligence tests, however, the same data were collected by mail for these subjects as for the in-state group seen personally.

The statistical treatment and analysis of the information collected in the follow-up was such an enormous and time-consuming task that it was decided in 1955 to bring the demographic information up to date for inclusion in the published report of the status of the gifted group at mid-life. Accordingly a two-page Information Blank was mailed to the subjects in the spring of 1955, calling for the latest information on the main items of basic data.

In addition to the data blanks* for the subjects and spouses listed in Table 3, Stanford-Binet tests have been given to a total of 1,525 off-spring of the gifted subjects. For a large proportion of these children a record of developmental history (Information About Child) was filled out by a parent, usually the mother, at the time of testing. The data from this blank will be reported in a separate publication at a later date.

TABLE 3

DATA SECURED FOR SUBJECTS AND SPOUSES, 1950–1955

	N	
	Subjects	Spouses
I. *1950–52 Follow-up*		
General Information (for subjects)	1,268	...
Supplementary Biographical Data (for subjects and spouses)	1,119	...
Data on Rate of Reproduction and Happiness of Marriage (for subjects)	972	...
Happiness of Your Marriage (for spouses)	...	565
Concept Mastery Test (subjects and spouses)	1,004	690
II. *1955 Follow-up*		
Information Blank (for subjects)	1,288	...

STATUS OF THE GIFTED STUDY 1950–1955

When the follow-up data of 1950–52 were gathered, the subjects had been under observation for approximately 30 years, and at the time of the 1955 survey the time had extended to about 34 years. Of the original group of 1,528 subjects, 91 had died by 1950 and an additional 13 deaths by 1955 brought the total number of deceased subjects to 104. During the course of study we have completely lost track of only 28 subjects (11 men and 17 women). In none of these cases has any contact been effected since 1928 or earlier, either with the subjects, their parents, or any members of their families. Although it is possible that not all of the "lost" subjects are now living, there is no reason to believe that our original loss of contact was in any case caused by the death of the subject but rather by the removal of the family to a different area. The eight-year interval between the field follow-up of 1927–28 and the mail follow-up of 1936 made it difficult to trace the subjects who had

* These blanks are reproduced in the Appendix.

moved any distance, especially since the relationships with the parents, and to an even greater extent with the subjects, had not been sufficiently well established by 1928 to ensure their notifying the research administration of changes in address. The greater loss in the case of women can be accounted for by the greater difficulty in tracing them because of name changes through marriage.

By 1955 death had reduced the number of subjects under study to 795 men and 629 women. The 11 men and 17 women classified as "lost" bring the number for whom information on current status was sought to 784 men and 612 women. In addition to the "lost" subjects there are a few other persons for whom our information is fragmentary, often secured indirectly through parent, sibling, or other informant. Furthermore, not everyone, including some of the most interested and co-operative, completed all the blanks; in a few cases, even, no blanks were filled out, the subject preferring to give the information in an interview or informally in a letter. Occasionally, too, a subject omitted one or more items in filling out a questionnaire. For these reasons there will be slight irregularities in the number of cases (N) for whom various types of data are reported in succeeding chapters. In the case of the Concept Mastery test the number of individuals is necessarily limited to those to whom the test could be administered in person. All subjects who were personally interviewed or for whom information was supplied either in a questionnaire or through correspondence are considered to have co-operated in the follow-ups.

The success of these two follow-ups is indicated by the almost incredible amount of co-operation that was secured. Of the 1,437 subjects living at the time of the field study, 95 percent participated actively and the addition of those for whom information was secured indirectly brings the total contacted, either directly or indirectly, to 97.5 percent. The results of the mail follow-up of 1955 are almost as impressive, with co-operation from 93 percent of the 1,424 subjects then living.* The follow-up contacts and co-operation for both the field study of 1950–52 and the mail follow-up of 1955 are shown in Table 4.

Scope of This Report

In the follow-ups of 1950–55 reported here, as throughout the study, the information obtained has not been limited to that secured by tests

* The 28 "lost" subjects are included in the totals of 1,437 and 1,424 since we are continuing to look for them.

TABLE 4
FOLLOW-UP CONTACTS AND CO-OPERATION, 1950–1955

	N	%of Total Group
I. *Field Study of 1950–52*		
Subjects interviewed by field workers	1,142	
Subjects contacted by mail only	225	
Total actively co-operating	1,367	95.1
No direct contact with subject but information from other sources	34	
Total in touch with directly or indirectly.....	1,401	97.5
No contact or information in 1950–52 but co-operating 1940–1945	8	
Lost : unable to trace and no information since 1928 or earlier	28	
Total group	1,437*	
II. *Follow-up by Mail, 1955*		
Information blank filled out by subject	1,288	90.5
No direct contact with subject but information from other sources	38	
Total in touch with directly or indirectly	1,326	93.1
Co-operating in 1950–52 but no information supplied in 1955	63	
No contact or information in 1955 but co-operating 1940–45	7	
Lost : unable to trace and no information since 1928 or earlier	28	
Total group	1,424*	

* The difference in the total group N's is accounted for by 13 deaths between 1952 and 1955.

and questionnaires. The data furnished by the subjects have been illuminated by the field worker reports, and much additional interesting and valuable information has come through informal personal correspondence and visits of the subjects or members of their families with the research staff at Stanford University. Such correspondence and visits have been frequent over the years, regardless of whether a follow-up was in progress.

This study is unique in many aspects, particularly in the length of time—almost 35 years—that the same group of individuals has been followed, and in the wealth of material collected about the subjects from childhood or early youth to mid-life, thus furnishing a continuous record of intellectual development and of educational, vocational, and marital history as well as of physical and mental health. In addition to these specific items there is a great deal of less easily statisticized data. These include the explorations into the personality dynamics, interests, and attitudes of the subject that complete the total picture of the gifted individual. The case history material is further enhanced by collateral data on the parents, the siblings, and the offspring. Finally, the unparalleled co-operation of the subjects and their families lends additional importance and validity to the findings.

So vast is the amount of information collected in the 1950–55 follow-ups that to discuss and evaluate it fully in this volume would delay publication until some of the items, particularly those related to vital and social statistics, would be out of date. This report, therefore, will be concerned chiefly with the data called for in the 1950 General Information Blank and the 1955 Information Blank. As a rule, in the case of those demographic items for which the information was secured at both dates, only the figures for the more recent date (1955) are given. Other findings to be covered in this volume include the results of the testing of the subjects and their spouses with the Concept Mastery test and of their offspring with the Stanford-Binet test. The tremendous amount of valuable autobiographical material supplied in the Supplementary Biographical Data blank as well as the data on factors affecting fecundity in the group as found in the Rate of Reproduction blank and on marital happiness as reported in The Happiness of Your Marriage blank will be only touched on in this volume. The detailed analysis and evaluation of the information from these questionnaires is reserved for future publication.

CHAPTER IV

MORTALITY, HEALTH, AND GENERAL
ADJUSTMENT

In addition to our interest in how gifted children turn out from the standpoint of educational, occupational, intellectual, and creative achievements, we also want to know about their careers as *people*. In this chapter we will consider the mortality record, physical health status, and the mental health and general adjustment of the gifted subjects. The latter topic includes, in addition to the ratings on general adjustment, such specific aspects of malfunctioning as mental disease, alcoholism, crime and delinquency, and problems related to sex.

MORTALITY

By 1955 the number of deaths among the gifted subjects was 104 (62 males and 42 females) ; this represents an incidence of 7.3 percent for males, 6.3 percent for females, and 6.9 percent for the total group of 1,500 subjects with whom we have been able to keep in touch. Table 5 gives the mortality rate according to sex and age at death.

TABLE 5

MORTALITY TO 1955
ACCORDING TO SEX AND AGE AT DEATH

Age at Death	Males (N = 846)		Females (N = 654)		Both Sexes (N = 1,500)	
	N	%	N	%	N	%
Under 15 years	3	0.4	2	0.3	5	0.3
15–24 years	22	2.6	14	2.1	36	2.4
25–34 years	19	2.2	14	2.1	33	2.2
35–44 years	17	2.0	12	1.8	29	1.9
45–49 years	1	0.1	..	...	1	0.1
All ages	62	7.3	42	6.3	104	6.9
Median age at death	29.5 years		27.6 years		28.4 years	

In comparing the mortality record of the gifted group with the rate for the generality the life table data supplied by Dublin and his associates for two different periods has been used. The first tables are based on the mortality conditions of 1929–31[12] and the second on the mortality conditions of 1939–41.[13] The data for these two periods reflect fairly well the conditions to which the gifted subjects have been exposed since only 14 of the total group selected in 1921–22 had died before 1929. The mortality rate for the generality in a particular age span can be determined from the life table by finding the number out of an arbitrarily large number (in this case 100,000) of live births who are living at a given age and the proportion of that number who die by a specified older age. Ages 11 and 44 were chosen as the initial and upper age limits, respectively, on the life table because they approximate the *average* ages of our group when first selected for study (1921–22), and at the time of latest report (1955). Under the conditions of 1929–31, the life tables indicate that of the United States white population who survive to 11 years of age, 12.7 percent of males, 10.8 percent of females, and 11.7 percent of the total cohort will have died by age 44. Under the improved mortality conditions of 1939–41 we find that in the general population 9.1 percent of males, 6.8 percent of females, and 8.0 percent of the total white population alive at age 11 will have died by age 44. A comparison of these figures with those given for the gifted subjects in Table 5 shows the mortality rate for our total group as well as for the sexes separately to be lower than the expectation based on conditions at either of the life table dates. This difference may be attributable to the superior physique and health that characterized the group in childhood as well as to their generally superior intellectual and economic status and its concomitants. However, as will be seen, accident-induced mortality has also been somewhat less than for the generality.

Causes of death. The causes of death with the percentage incidence are given in Table 6. Natural causes account for 62 of the 104 deaths. Accidents follow with 21 deaths, and suicide ranks third with 15 deaths. The five World War II casualties are listed separately. Three of these men died in combat, one was killed in the crash of a transport plane he was piloting on military duty, and a naval officer lost his life in a storm at sea.

Among the natural causes the three leading diseases have been: the cardiovascular-renal group with 17 deaths, cancer (including 3 cases of leukemia) with 10 deaths, and tuberculosis with 9 deaths. All but four of the 27 deaths from the first two causes occurred in the 15-year

TABLE 6

CAUSES OF DEATH

	Males (N = 846)		Females (N = 654)		Both Sexes (N = 1,500)	
	N	%	N	%	N	%
Natural causes	30	3.5	32	4.9	62	4.1
Accidents	17	2.0	4	0.6	21	1.4
World War II casualties	5	0.6	..	...	5	0.3
Suicides	10	1.2	5	0.7	15	1.0
Cause not known	..	...	1	0.1	1	0.1
Death from all causes ...	62	7.3	42	6.3	104	6.9

period from 1939 through 1954. This agrees with the pattern in the general population of an increase in the death rate from these diseases with advance in age. On the other hand, with one exception, all the deaths from tuberculosis occurred before 1939 and therefore among younger subjects.

For gifted men, only the cardiovascular-renal diseases among the natural causes have an incidence as high as the death rate from accidents; each has caused the death of 17 men. Among gifted women three diseases rank higher than accidents as a cause of death: tuberculosis and cancer have taken six lives each, and diseases of the heart account for five deaths. In the general population, accidents rank first (ahead of any single disease or defect) as a cause of death from birth to 45 years for males and to 25 years for females, and are in fourth place for the total population of all ages. Of the accidental deaths those caused by motor vehicle accidents were the most frequent in this group, just as they are in the general population. Eight of the 17 accidental deaths among the males and all four of those among females resulted from automobile accidents. Of the other 9 men, 4 were killed in airplane crashes and 2 died in industrial accidents. One death resulted from each of the following: drowning, accidental gunshot, and a fall while mountain climbing.

Suicide, the third most frequent cause of death in the group, arouses particular interest because of the tragedy attached to these deaths and the possibility that with help at the proper moment some of them at least could have been avoided. Data for accurate comparison of the incidence of suicide in the gifted group with that in the generality are lacking. The vital statistics reports give only the incidence of suicide at a given time—in other words, a "snapshot" picture of the situation. The prevalence rates mentioned in most of the research studies con-

cerned with the personality and motivations of the individual who com-
mits suicide are also cross-sectional, and are limited to a particular
date or period of time.

The most helpful information on the amount of suicide in the population at large comes from Dublin and his associates,[13] who have computed the chances per 1,000 at decennial ages of eventual death from
certain specified causes. These expectancies are based on the life table
and death data for the United States during 1939–41. Dublin[10] points
out that the over-all picture of suicide in the United States shows practically no change in rate during the past 50 years, though there have
been some minor fluctuations, with the highest rates occurring in 1932
at the bottom of the depression and the lowest rates during wartime.
Dublin's 1939–41 data can therefore be considered representative of
the trend of suicide during the lifetime of our group. According to the
Dublin data, the chances per 1,000 population, age 10 to 19 years, of
eventually dying by suicide are 17.5 for white males and 5.5 for white
females. In terms of percentages, the expected incidence of suicide
among the total population in this age interval is approximately 1.8
percent for males and about 0.6 percent for females. As shown in
Table 6 the suicide rate in the gifted group to 1955 is 1.2 percent for
men and 0.7 percent for women.

Comparison of the data for the gifted with that for the total population should take several factors into account. First, the figures for a
life table population are derived statistically and, while they serve very
well for generalizations, are less applicable to small groups than are
data based on an actual population. Especially in a group such as ours,
which is not only relatively small but also heterogeneous as regards
date of death and age at death, differences from the generality may well
be caused by chance. In considering the suicide rate, still other aspects
of the situation should be borne in mind. Among these is the well-
known and acknowledged fact that all quantitative analyses of suicide
in the generality are underestimates since many suicides are never so
reported. Not only may the person committing suicide succeed in
making his death appear natural or accidental, but also the relatives
and friends often take pains to conceal the fact that death was self-
inflicted. In the case of our gifted subjects we have what we believe
to be full information regarding the cause of death for all but one subject. This was a woman who had at one time attempted suicide but
who was also in poor health for a number of years, and the report of

her death received from a former employer did not suggest suicide. Other factors to be considered in appraising the amount of suicide in the gifted group are the regional and socio-economic differences in suicide rate found for the generality. Semelman[30] reports that suicide is most frequent in the West, especially in California, with San Francisco having one of the highest rates in the nation. Henry and Short,[18] basing their conclusions not only on their own studies but also on a survey of the work of other investigators, point out the positive relationship of suicide to status, both social and economic. Dublin finds that suicide is more common in urban than in rural populations and more frequent among white than among colored people. On the basis of these findings, it appears that the gifted subjects as a group possess the characteristics that make for a higher suicide rate: Californian in origin, with approximately four-fifths still residing in the state; chiefly urban; white; and of superior status from the standpoint of education, occupation, income, and achievement.

Bearing these factors in mind, one hesitates to draw any clear-cut conclusions from a comparison of the suicide rate in our group with Dublin's expectancy rate. However, on the surface at least, the rate for the gifted appears to be high, especially in the case of women who have already exceeded the total expectancy given by Dublin: 0.7 percent for gifted women as compared with 0.6 percent for all women. The gifted men are still below the estimate of eventual suicide in a 10- to 19-year-old cohort: 1.2 percent for the gifted as compared with 1.8 percent for the generality of males. Before closing this discussion, mention should be made of the discrepancy in sex differences in suicide rate between the gifted group and the total population. All investigators report a much higher frequency of suicide among men, ranging from 3 to 4 times the incidence among women. Among the gifted subjects, however, less than twice as many men as women, proportionately, have committed suicide. One cannot, of course, overlook the possibility of a chance difference here owing to the relatively small numbers involved.

Subjects who have died. The average age at death for those who have died was approximately 28 years with a range from 8 to 45 years (see Table 5). Their IQ's ranged from 134 to 184 with an average of 149. Ten of the deceased were in elementary or high school and 17 were undergraduate students in college at the time of death. Another 5 were graduate students. Among those who had completed their schooling or were at the graduate student level, 50 percent of the men and 38 percent of the women had taken a bachelor's degree and 35 percent of men and 18 percent of women had attended college for from one to four years. The graduate degrees taken include 3 Ph.D.'s, 3 M.D.'s, 3 LL.B.'s, and 5 master's. In addition, two of the graduate students among the deceased men were candidates for a master's degree and the third was working toward an engineering degree. Both women graduate students were on the point of getting a Ph.D., one in astronomy and one in psychology. Three of the men were physicians, 3 were lawyers, and 2 were members of university faculties. Other occupations represented among the deceased men were chemist, business executive, musician, advertising and public relations work, mechanic, salesman, etc. The majority of the women who died after completing their education were married and occupied as housewives. However, one was a college teacher, one a statistician, one a librarian, and one a high-school teacher.

General Health and Physique

Information on general health, based on interviews with the subjects and often with the spouses as well, was supplied by the field workers in the 1939–40 and the 1950–52 follow-ups. In addition, self-ratings by the subjects on physical health were called for in the 1940, 1950–52, and 1955 information blanks. Table 7 gives the ratings on

TABLE 7

SELF-RATINGS ON PHYSICAL HEALTH

	Men			Women		
	1940 (N = 700)	1950 (N = 750)	1955 (N = 714)	1940 (N = 563)	1950 (N = 601)	1955 (N = 567)
Self-rating	%	%	%	%	%	%
Very good	52.3	50.5	46.5	44.7	43.3	43.0
Good	38.6	41.6	44.8	39.0	43.4	45.5
Fair	7.3	6.2	7.0	12.7	11.1	8.3
Poor or very poor	1.9	1.7	1.7	3.7	2.2	3.2

physical health for the three follow-up dates when the subjects were at the approximate average ages of 29, 40, and 44, respectively. These are substantially self-ratings, although in a few instances it was necessary to interpret or modify a self-report in the light of the case history evidence. However, such modifications were felt to be warranted in less than one percent of the cases.

It will be noted that the ratings are consistent for each of the three

reports, with more than 90 percent of the men and from 84 to 88 percent of the women rating their health as good or very good. The somewhat lower rating on general health for women is a characteristic sex
difference that has been found also for men and women in general.[7]
Statistics show that among females as a whole morbidity is higher, while
among males the mortality rate is higher. We have already seen in
Table 5 that the mortality rate for the gifted men is higher at each age
than that for the gifted women.

The questionnaires of 1950–52 and 1955 did not ask for information on height and weight. However, the medical examinations and
anthropometric measurements that were made following the selection
of the subjects in the original survey showed the gifted children as a
group to be above the best standards for American-born children in
growth status as indicated by both height and weight. The records
showed that they were also above the established norms for unselected
California children.[39] According to the information supplied by the
subjects in 1940, the median height of the adult gifted men was 71.3
inches and of the adult gifted women, 65.2 inches. The average weight
of the men was 162.8 pounds and of the women, 126.6 pounds. After
making allowance for some overstatement in the self-report, it was
estimated that the gifted men average about one-half inch taller than
college men in general and about one and one-half inches taller than
the generality of men in the United States of their generation. Similarly, gifted women average close to one-half inch taller than college
women in general and approximately one inch taller than women in the
total population. The relationship of weight to height appears to be
about the same in the gifted as in the total population.[36]

Physical defects sufficient to handicap the individual seriously are
infrequent among our subjects. Nine men are crippled, but only one
case is sufficiently serious to be disabling; for most, the impairment is
no more than a slight limp. In seven cases the crippling resulted from
poliomyelitis and in the other two cases it was due to a congenital defect.
Among the women there are seven cases of orthopedic handicap. For
two of these the difficulty is a congenital hip dislocation which resulted
in some crippling but not severe enough to prevent independent loco-
motion. Poliomyelitis has left three other women with slight muscular
impairment but not greatly handicapped. More seriously afflicted are
two women who are confined to wheel chairs, one since childhood because of crippling arthritis and one as a result of poliomyelitis contracted

at age 39. The former, who holds a Ph.D., is a distinguished scholar, writer, and university professor. The latter, a lawyer with a well-established practice (and also a housewife and mother) when disabled, resumed her career as soon as possible and is continuing her law practice besides managing her home from her wheel chair.

MENTAL HEALTH AND GENERAL ADJUSTMENT

The gifted subjects have been rated on mental health and general adjustment at various stages of follow-up. The most important sources of information have been the personal conferences by the research staff with the subjects, their parents, and their spouses; letters from the subjects or members of their families, or other qualified informants; and responses to questionnaires filled out by the subjects and, in the earlier years, by their parents also. The information schedules in the follow-up surveys of 1936, 1939–40, 1945, 1950–52, and 1955 all included the question: "Has there been any tendency toward nervousness, worry, special anxieties or nervous breakdown in recent years? . . . nature of such difficulties . . . ." From 1940 on, this item was supplemented by a question on how the difficulty, if any, was handled and on the present condition of the subject. Each specific kind of information obtained was considered in the light of total case history.

With these accumulated data at hand the mental health and general adjustment of each subject was assessed and the subjects classified according to three categories as follows: 1, satisfactory adjustment, 2, some maladjustment, and 3, serious maladjustment. The third category was divided into two sub-groups: 3a, serious problems in adjustment but not severe enough to require hospitalization, and 3b, a history of hospitalization for mental illness. These categories correspond to those used for the 1940 and 1945 ratings[36] and are defined as follows:

1. *Satisfactory.* Subjects classified in this category were essentially normal; i.e., their "desires, emotions, and interests were compatible with the social standards and pressures" of their group.[23] Everyone, of course, has adjustment problems of one kind or another. Satisfactory adjustment as here defined does not mean perfect contentment and complete absence of problems but, rather, the ability to cope adequately with difficulties in the personal make-up or in the subject's environment. Worry and anxiety when warranted by the circumstances, or a tendency to be somewhat high strung or nervous—pro-

vided such a tendency did not constitute a definite personality problem —were allowed in this category.

2. *Some maladjustment.* Classified here were subjects with excessive feelings of inadequacy or inferiority, nervous fatigue, mild anxiety neurosis, and the like. The emotional conflicts, nervous tendencies, and social maladjustments of these individuals, while they presented definite problems, were not beyond the ability of the individual to handle, and there was no marked interference with social or personal life or with achievement. Subjects whose behavior was noticeably odd or freakish, but without evidence of serious neurotic tendencies, were also classified in this category.

3. Serious maladjustment.

a) Classified as 3*a* were subjects who had shown marked symptoms of anxiety, mental depression, personality maladjustment, or psychopathic personality. This classification also includes subjects who had suffered a "nervous breakdown," provided there had not been a mental disorder of sufficient severity to require hospitalization. Subjects with a previous history of serious maladjustment or so-called nervous breakdown were included here even though their adjustment at the time of rating may have been entirely satisfactory.

b) Classified as 3*b* were those subjects who at any time had suffered a complete mental breakdown requiring hospitalization, whatever their condition at the time of rating. In the majority of cases the subjects were restored to reasonably good mental health after a brief period of hospital care.

[NOTE: Attention is called to the fact that ratings 3*a* and 3*b* are historical in character; i.e., if a person has *ever* been seriously maladjusted as defined above, the 3 rating (*a* or *b*, according to degree) continues to be used even though he may now be greatly improved or even free from difficulty.]

Results of ratings. In 1955 there was sufficient information on hand to rate the mental health and general adjustment of approximately 98 percent of the 1,396 living subjects with whom contact had been maintained.* For the remaining 21 men and 11 women some

---

* The 1955 ratings are a composite of the data collected in the 1950–52 field follow-up and the 1955 follow-up by mail. Since the 1950–52 data are more extensive, they formed the basis for the rating. In 1955 these ratings were brought up to date, and any changes in status are shown in the ratings presented here. The 28 subjects who had been "lost" since 1928 or earlier are not included in the total of 1,396 subjects.

information was available but it was felt the data were not sufficiently complete to permit a definitive rating. However, on the basis of what was known about these subjects and their activities, there is no reason to believe that any had been hospitalized for mental disorder. Part I of Table 8 gives the ratings for the subjects who were living and for whom data on mental health and general adjustment were adequate; in Part II of the table the remainder of the original group is accounted for.

TABLE 8

RATINGS ON MENTAL HEALTH AND GENERAL ADJUSTMENT

	Men		Women	
	N	%	N	%
Part I. Rating				
1. Satisfactory	525	68.8	396	65.9
2. Some maladjustment	170	22.3	151	25.1
3. Serious maladjustment				
a) Without mental disease*	47	6.2	36	6.0
b) Hospitalization for mental illness..	21	2.7	18	3.0
Total rated	763		601	
Part II. Not Rated in 1955				
A. Information not complete enough for rating, but no record of hospitalization for mental illness	21		11	
B. Deceased before 1955	62		42	
C. No contact since 1928 or earlier and no information regarding status of subject	11		17	
Total group, living and deceased	857		671	

* Mental disease is defined as hospitalization for mental illness. (Cf. discussion on pp. 41–42.)

Of the 763 men and 601 women rated for general adjustment, better than two-thirds of the men and close to two-thirds of the women were considered satisfactory in adjustment. Somewhat more than one-fifth of the men and one-fourth of the women were rated in category 2 and the 9 percent of each sex who had experienced serious difficulty were classified in category 3. When this latter category is broken into subgroups, we find that 6 percent fall in the 3*a* group, that is, a history of serious maladjustment but without reaching the severity of a mental breakdown. The approximate 3 percent classified as 3*b* all had a history of hospitalization for mental illness. The percentages in the 3*b* category would be somewhat reduced if the 21 men and 11 women of

Part II-A of Table 8 were included in the total number. Although not enough was known about these additional cases to rate them on general adjustment, there is no evidence that any has been hospitalized for mental illness.

As indicated in the definitions, ratings of "some" or "serious" maladjustment were given on the basis of total case history data and do not necessarily represent the current status of the subject. This method of rating was used to show the extent to which our group had suffered from problems of personality or emotional adjustment regardless of whether the difficulty had been overcome. It is important, therefore, to view the ratings on adjustment in the light of the estimates of "present condition" called for in the Information Blank as a supplement to the question regarding nervous and emotional difficulties. Information on present condition was available for 235 men and 202 women, or all but 3 men and 3 women among those who were rated either 2 (some maladjustment) or 3 (serious maladjustment). Although self-estimates of this kind are subject to error, the replies were checked against the field worker reports and other case history data for verification and only occasionally was it necessary to modify the response of the subject. In the few cases in which the subject omitted a reply to the query on present condition, the information, if available from other sources, was supplied. The estimates of present condition for subjects rated 2 or 3 were distributed as follows:

	Percentages	
Present Condition	Men	Women
Free from difficulty	21	16
Improved	55	68
No change	20	12
Worse	4	4

It is especially interesting to find that of the 39 subjects living in 1955 who had undergone hospitalization for mental illness, 28 rate their present condition as improved, and 2 report themselves as free from difficulty. Another 4 of those rated 3b say there has been no change, and 5 rate their condition as "worse."

So far our discussion of general adjustment has been limited to subjects who were living in 1955 and who could, therefore, be rated on general adjustment. However, to get the full picture of the gifted group from the standpoint of mental health, account should be taken of the subjects who have died. In view of the wide range in date of death (1923 to 1954) and age at death (8 years to 48 years), it would not be practical to attempt to classify the deceased according to the mental adjustment categories; however, those with a history of hospitalization can be used in an extension of category 3b to include the deceased. Of the 104 subjects who had died by 1955, five men and four women had, at some time, been patients in a mental hospital. When the 62 deceased men and 42 deceased women are added to the number of each sex rated (Part I, Table 8), the results are as follows for the 825 men and 643 women, living and deceased, whose status could be evaluated:

	Men		Women	
	N	%	N	%
No record of hospitalization for mental illness	799	96.9	621	96.6
History of hospitalization	26	3.1	22	3.4

Subjects with a history of mental disorder. The figures of 3.1 percent for men and 3.4 percent for women include all subjects, whether presently living or deceased, who have ever been admitted to a hospital or sanitarium for the care of the mentally ill regardless of the seriousness of the illness or the length of the hospitalization. In most cases the illness was comparatively mild and the hospitalization brief, some less than three months, few more than a year. However, there have been 7 cases of prolonged hospitalization. These included 2 epileptics, both men, one of whom died in the hospital at the age of 30. He had suffered from epilepsy most of his life but the illness did not become disabling until he was about 20 years old. Before hospitalization became necessary he had completed one and one-half years of college work. In the case of the second man, the onset of epilepsy in recognizable form took place after college graduation just as he was embarking on a professional career. After a brief hospitalization he was able to take a clerical job and worked for some years. Apparently recovered, he returned to the university for a year of graduate work with the purpose of re-entering his chosen profession. However, a recurrence of the epilepsy sent him back to the hospital. The next few years saw him in and out of hospitals and working at simple occupations when able. Finally the illness became so severe that institutional care became necessary and he has now been hospitalized for nearly ten years.

A third man, hospitalized for three and one-half years as a schizophrenic, was recently released. While he has not fully recovered, it was felt that his improvement was sufficient to permit home care.

The 4 women with prolonged illness include one now age 40 who became ill when she was a university student and was sent to a private sanitarium where she has been for the past twenty years. The diagnosis was dementia praecox and at last report the outlook for ultimate recovery was poor. Another woman who had completed her education, married, and had 2 children, was stricken with encephalitis at the age of 31 with resulting brain damage so serious that she has been incapacitated ever since her illness nearly fifteen years ago. The third of the 4 women with prolonged illness presents a brighter picture. This woman, who is unmarried, completed college with an excellent record. Several years of successful employment followed until her first breakdown at age 29. After three mental breakdowns with brief hospitalizations, her condition became so much more serious that she was committed to a state hospital with a diagnosis of dementia praecox, paranoid type. During her twelve-year stay in the hospital she was able, with the exception of occasional brief periods, to do secretarial and library work with great efficiency. By 1954 her mental health had so improved that she was released from the hospital on a home-care basis. Her ultimate discharge will depend on how full and permanent her recovery proves to be.

The fourth woman requiring lengthy hospitalization went into a state hospital after six weeks in a private sanitarium. She was then 32 years old, single, with two years of college education, and had been employed as a secretary for a number of years. The diagnosis was dementia praecox, mixed type. After nearly three years in the hospital she was released only to have a recurrence three years later, which again necessitated hospitalization. A year later she was discharged and, so far as is known, has had no further relapses.

Except for alcoholism in the case of men, the manic-depressive states have been the most frequent type of disorder. The frequencies of the various kinds of mental illness classified according to primary diagnosis where available were as follows:

	Men N	Women N
Alcoholism	10	3
Epilepsy	2	..
Manic-depressive state	7	8
Psychoneurosis	5	4
Schizophrenia (dementia praecox)	2	6
Traumatic brain lesion	..	1

In all but one case (reported directly from the attending physician in a personal communication) our information on the nature of the illness has come from the subject or a close family member, and the diagnosis often but not always confirmed by medical authorities. In the hospitalized cases reported simply as "nervous breakdown" or similarly, we are especially uncertain of the nature of the illness and, lacking full information for these cases, we have classified them under the heading of psychoneurosis. The total number of hospitalizations is so small that percentages have not been computed for the various mental disease classifications, since they would not be comparable to the proportions in the various mental illness categories for hospital admissions in general. A comparison with the generality of admissions would also be distorted by the age range of our group. As one would expect, we have no cases of organic psychoses resulting from vascular disease characteristic of later life.

The ages at hospitalization (first admission if more than one) for the subjects, both living and deceased, with a history of mental illness were distributed as follows:

Age in Years	Men N	Women N
10–19	...	2
20–29	6	6
30–39	15	8
40–49	5	6

The IQ's of these subjects ranged from 140 to 180 and their schooling varied from high-school graduation to several years of university graduate work. All but 6 of the men had had some college work and 14 had graduated from college. The latter included 3 men with an M.D. and 4 with an LL.B. degree. Eleven of the women were college graduates and 3 had also taken a graduate degree. Another 8 women attended college for from one to three years and 2 had only a high-school education. In addition, there are 2 subjects (1 man and 1 woman) among those hospitalized who died before completing their education. Both had taken approximately two years of college work.

Comparison of mental disease among the gifted group with the rate for the generality. The criterion most widely used in studies of the frequency of mental disease in the United States is the number of admissions to hospitals for the care of the mentally ill, since that is the only objective yardstick available. The term "expectation of mental disease" is used to describe the chances of being hospitalized (first time, if more than once) in an institution, either public or private, for the care of the mentally ill. In contrast to the prevalence measures which give the proportion of a given population hospitalized at a specified time, the expectancy measure gives the cumulative probability of admission to a mental hospital. This is the sort of information needed for comparison with the rate of mental disease in the gifted group where the data are also cumulative.

Different methods of calculating the expectancy of mental disease yield somewhat different estimates, but careful investigations such as those of Malzberg[26] and of Goldhamer and Marshall[15] indicate that 8 to 10 percent of the U.S. population will be admitted to a hospital for the care of the mentally ill at some time during their lives.* Malzberg's expectancy tables, which follow the principle of the life table, show that of 1,000 population alive and sane at 10 years of age, 85.4 males and 85.9 females will develop a mental disease. Goldhamer and Marshall suggest a different method of calculation, which gives first admission expectancy for a member of a particular population group *if* he survives to a specified later age. This "conditional expectancy" rate is in contrast to the Malzberg data which give a joint expectancy, i.e., the combined probability of survival *and* of admission to a mental hospital. Both methods serve a purpose and their relative merit cannot be evaluated here. The Goldhamer-Marshall conditional expectancy rates, however, are more appropriate to our data and so will be used in the following comparisons of the gifted with the generality.

According to the conditional expectancy tables, of those who survive to the age of 75, about 1 in 10 persons will be hospitalized. More pertinent to the current data for our gifted group than lifetime expectancies of mental disease are the chances of hospitalization to mid-life. This information also is available in the Goldhamer-Marshall tables which give the conditional expectancy between specified initial and later ages. For comparison with the data for the gifted, we have chosen the initial age of 10,† which is near the average age of the subjects when selected

---

* Malzberg, and Goldhamer and Marshall have based their calculations of the expectancy of mental disease on the 1940 first admission rates to state and licensed institutions, public or private, in New York State. It is believed that these figures would agree fairly closely with the expectancy rates in those areas where adequate facilities are available for the care of the mentally ill. California, in which 80 percent of our subjects reside, is one of these.

† Since the Goldhamer-Marshall data are reported for quinquennial ages, 10 years is the nearest to the average age (about 11 years) of our group when selected for study.

for study and the terminal ages of 40, 45, and 50. These later ages approximate the age distribution of the gifted subjects in 1955 when the median age was about 44 years. Although the actual age range in the gifted group was about 20 years, 84 percent were born between the years 1905 and 1915, inclusive, and so were between the ages of 40 and 50 in 1955. According to the conditional expectancy table the chances in 100 of admission to a mental hospital between age 10 and the specified later ages for the general population are as follows:

Terminal age	Percentage of General Population	
	Male	Female
40	2.8	2.4
45	3.4	3.0
50	4.2	3.7

The proportion of gifted subjects, including the deceased, who had been hospitalized for mental illness up to an average age of about 44 years (1955) was 3.1 percent for men and 3.4 percent for women. A comparison of these figures with the Goldhamer-Marshall data shows the incidence among gifted men to be slightly below the expectancy for the male population of comparable age, and among gifted women, slightly above the expectancy. Probably neither sex differs greatly from the generality in the frequency of mental disease. One should not overlook the possibility that the rate of hospitalization among the gifted may be related to some extent to their generally superior status; not only intellectual—which may give them insight into their needs, but also socio-economic—which makes it possible for them to seek aid. The majority of the hospitalizations have been voluntary and a large proportion have been in private institutions. On the basis of Malzberg's more than 8 percent expectancy of eventual mental disease and the Goldhamer-Marshall approximate 10 percent expectancy to age 75, it is questionable if the incidence of mental disease in the gifted group will exceed the expectancy for the generality, particularly because of certain inverse relationships that have been found between educational-social-economic status and mental disease rate in later maturity.[28] To equal the expectancy, the present incidence of slightly over 3 percent in our total group would have to be almost tripled. Whether the extent of mental disorder reaches that proportion remains to be seen.

Suicide and mental disorder. As reported earlier in this chapter, 10 men and 5 women had committed suicide by 1955. Of these 15 subjects, one man and two women had been hospitalized for mental illness, and only these three cases among the suicides are considered to have a history of mental disease as we have defined it. There is no doubt that death forestalled eventual hospitalization of additional persons among the suicides. In other cases, however, although there had been indications of maladjustment, the difficulties had not appeared to be serious enough to constitute mental disease and in two cases, both women, there had been no evidence of a serious adjustment problem. Opinion is divided among psychiatric authorities as to whether it is only the psychotic individual who takes his own life. Though many leading psychiatrists hold this opinion, there are others who believe that the "sane" individual also may commit suicide. Dublin and Bunzell,[11] whose study of suicide is one of the best in the field, feel that in view of the conflicting psychiatric opinions it is extreme to assert that all suicides are insane—unless it is assumed a priori that self-destruction is in itself a definite indication of psychosis. These authors do not concur in such an assumption, but believe, rather, that suicide does occur among individuals who should be designated as sane, even though it is a far greater hazard among sufferers from mental disease.

Whether or not suicide is considered a psychotic manifestation per se does not affect our comparison of the incidence of mental disease in the gifted group with that in the generality since, in both cases, mental disease is defined as admission to a hospital for the treatment of the mentally ill.

Use of liquor among the gifted subjects. Our records show that for 10 men and 3 women the precipitating cause of hospitalization in a mental institution was alcoholism, regardless of whatever underlying personality disorders there may have been. Furthermore, among those hospitalized for functional psychoses, 2 men and 1 woman had a serious alcohol problem as well, although this was not the primary reason for hospitalization. In addition to the information on alcohol as a problem among the subjects with a history of mental disease, considerable data have also been obtained on the extent to which alcohol is used by the group as a whole. The General Information Blanks of 1940 and 1950 included a specific question on this matter calling for the individual to rate himself according to several categories on the use of liquor. In 1950 this item was presented in the following form:

Use of Liquor. (Check the statement below that most nearly describes you)

———I never take a drink, or only on rare occasions.

———I am a moderate drinker. I have seldom or never been intoxicated.

————I am a fairly heavy drinker; I drink to excess rather frequently but do not feel that it has interfered seriously with my work or relationships with others.

————Alcohol is a serious problem. I am frequently drunk and attempts to stop drinking have been unsuccessful.

A questionnaire never provides for all the exceptions and variations found in the responses given by the individuals. From the comments of the respondents to this item, it was possible to add a fifth category to the four presented in the information blank. This category included those with a history of excessive drinking who no longer drink at all or who drink only moderately. The classification of the subjects according to use of liquor was based on a composite of the self-ratings, the field worker reports, and other pertinent case history information. The results of the ratings on use of liquor (as of 1950–52) are given in Table 9.

TABLE 9

RATINGS ON USE OF ALCOHOLIC DRINKS

	Men		Women	
	N	%	N	%
A. Never, or only rarely, take a drink	127	16.9	194	32.4
B. Moderate drinker	520	69.0	369	61.7
C. Fairly heavy drinker	87	11.6	29	4.9
D. Alcohol is a serious problem	10	1.3	3	0.5
E. Formerly a serious problem, now under control	9	1.2	3	0.5

It is interesting to find that about 17 percent of men and 32 percent of women never, or only on rare occasions, use alcoholic drinks and that another 69 percent of men and 62 percent of women describe themselves as moderate drinkers. About 12 percent of men and slightly more than 5 percent of women are classified as heavy or problem drinkers. Persons classified in categories D and E of Table 9 were considered to be, or to have been, alcoholics in the sense of the World Health Organization definition of alcoholism.* On the basis of this definition, 10 men and 3 women were rated as alcoholics in 1950–52. Another 9 men and 3 women had been problem drinkers in the past but were now able to

* The definition adopted by the World Health Organization in 1951 is as follows: "Alcoholics are those excessive drinkers whose dependence upon alcohol has attained such a degree that it shows a noticeable mental disturbance, or an interference with their bodily or mental health, their interpersonal relations, and their smooth social and economic functioning; or who show the prodromal signs of such developments."

control the difficulty. The E category of Table 9 includes 3 men who had been hospitalized for alcoholism. While others in this category received psychiatric or psychological aid, the majority attribute their success in overcoming excessive drinking to Alcoholics Anonymous.

The data in Table 9 represent the status of the group at a particular period of time in contrast to the mental health ratings which are historical in nature. It is possible, therefore, to compare the current extent of alcoholism in our group with figures for the generality. Jellinek and Keller[20] report that in 1948 the rate of alcoholism in the United States, as defined by the World Health Organization, was 3,952 per 100,000 adult population (roughly 4 percent). The United States rate was also computed separately by sex and showed an alcoholism rate of about 7 percent for men and slightly more than 1 percent for women. The highest rates, according to these investigators, are found in Nevada, California, and New York. In California, where approximately four-fifths of the gifted group reside, the 1948 alcoholism rate was 6,888 per 100,000 adult population (sexes combined). Separate figures according to sex were not reported for the individual states, but on the basis of these figures it is clear that the California rate for each sex is considerably higher than the 7 percent of men and 1 percent of women alcoholics reported for the total U.S. population. In contrast, only about 1 percent of gifted men and one-half of 1 percent of gifted women were classified as alcoholics.

Crime and delinquency. As shown in our earlier report,[36] the incidence of crime and delinquency is very low. Three subjects (all boys) had youthful records of delinquency that resulted in their being sent to a juvenile reformatory. In addition, one man served a term of several years in prison for forgery. All four of these are married, employed, and fulfilling their duties as responsible citizens. Three other boys came before the Juvenile Court for behavior difficulties but after brief detention were released to their parents. Among the gifted women only two are known to have had encounters with the police. Both were arrested for vagrancy and one served a jail sentence. Although each of these women has a history of several marriages, both seem to have become much more stable in recent years, and to have made normal behavioral adjustments.

Problems of sex. Information on sex problems is available from the various case history reports and information schedules supplied by subjects, parents, and field workers over the years of follow-up. Espe-

cially pertinent were the questionnaires on personality and temperament (1940) and marital adjustment (1940 and 1950), and particularly
the Supplementary Biographical Data blank filled out in the 1950–52
follow-up. This blank included a direct question regarding sex problems, worded as follows: "Either in childhood or later have there been
any major problems or marked difficulties related to sex?"

Of those who filled out this blank, 78 percent of men and 77 percent of women indicated the absence of any serious sex problems. The
problems mentioned by the remaining 22 or 23 percent of the group
covered a wide range over both the nature and the gravity of the difficulty. The most frequent problem was that of sexual adjustment in
marriage, mentioned by 4.6 percent of men and 6.5 percent of women.
For approximately the same number of men (4.5%), the problem involved shyness, awkwardness, fear of failure, or rebuff in relations with
the opposite sex, but less than 2 percent of women mentioned this problem. Aversion to sex or feelings of guilt constituted a sex adjustment
problem for 4 percent of women and 1 percent of men. Masturbation
was mentioned as having been a source of difficulty (chiefly in childhood and adolescence) by 4 percent of men and 2 percent of women.
About 2 percent of both men and women reported concern about a high
degree of sex drive (considered it above average) while an equal number of men and somewhat more women (3.5%) felt that their problem
was a lack or decline of sex drive and interest. Other factors mentioned
as making adjustment difficult were lack of sex education (less than 1%
of men and slightly more than 1% of women) and early sex experience
or sex shock in childhood or youth (0.3% of men and 1.3% of women).
Except for homosexuality, the other problems listed were relatively
minor and none was cited by more than 1 percent of either sex.

The sex problems so far discussed, though presenting difficulty for
the individual, have not been of the dimensions to constitute an aber-
ration or an insuperable obstacle to adjustment for the individual.
Homosexuality, on the other hand, is a deviation of such serious proportions involving both personal and social adjustment that its incidence in the gifted group has been reserved for separate discussion.
Our concern here is with subjects who have had homosexual experiences and for whom heterosexual adjustment has been difficult or impossible. Homosexuality, thus defined, has been reported for 17 men
(2%) and 11 women (1.7%). Undoubtedly, there are, in addition,
instances of latent homosexuality in which there has been no overt ex-

pression or even recognition of the tendency. Estimates of the prevalence of homosexuality in the general population vary, but all authorities put the incidence well above that found in our study. The most recent and best known investigation of sex behavior is that of Kinsey and his associates[21] who report that 25 percent of the male population between ages 16 and 55 have more than incidental experience and that 4 percent of white males are exclusively homosexual throughout their lives. Kinsey finds homosexuality much less common among females with an estimated frequency of from one-half to one-third of that reported for males. The figures for both sexes in the general population give an incidence many times that found for the gifted subjects. In contrast to Kinsey's report, that homosexuality occurs among 25 percent of males and 8 to 12 percent of females, only 2 percent of gifted men and 1.7 percent of gifted women are known to be presently homosexual or to have had homosexual experiences. It is possible that a very few cases have escaped our notice; however, our data are so complete that these few, if there are any, would not increase the incidence of homosexuality significantly.

Ten, or somewhat more than half, of the 17 gifted men classified as homosexual are exclusively so. These include 9 who are overt homosexuals, and except for one case, none of these men has ever married. In the one case, the marriage was of short duration and ended in divorce. Another man, well aware of his basic homosexuality, has abstained from overt manifestation. He has found an outlet in a highly successful career in the arts, and appears to have achieved an effective sublimation of his homosexuality. Six of the 17 men with a homosexual history have married and have made a reasonably satisfactory heterosexual adjustment. The remaining man, though he had not married, was reported to be making a successful transition to heterosexuality when he was killed in an accident.

All but one of the 11 women in the gifted group who have had homosexual experiences have married, and six of the marriages are apparently successful. In two cases the marriage broke up within a very short time and the women resumed a homosexual pattern of life. Another woman has had three unsuccessful marriages and at last report was living as a homosexual. One of the women who was married briefly has long been a mental patient and is reported by the hospital as an overt homosexual. The remaining woman has never married but has had heterosexual as well as homosexual experiences.

There is no doubt that homosexuality has interfered with the personal and social adjustment of these persons. For four men and two women, alcohol became a serious problem, resulting in hospitalization for two of the men and both women. On the other hand, at least 5 of the male homosexuals are rated among our vocationally most successful men.

Relation of general adjustment to education and intelligence. Because only about one-half of those who have suffered from mental disease are college graduates in contrast to the nearly 70 percent of the total group who completed college, a relationship between general adjustment and amount of education among the gifted might be suspected. In the case of men, at least, this indication is not borne out by the complete record. A comparison of the extent of schooling with general adjustment rating shows practically no difference for men at three levels of education. Actually, what difference there is favors the men who did not go to college. The college graduates and the group who attended college one or more years without graduating have almost precisely the same proportion of satisfactory ratings. For the women, on the other hand, there is evidence that the present status of those who did not go to college or who entered but did not complete college is less satisfactory with respect to general adjustment than that of the college graduates. Table 10 compares the ratings on general adjustment according to educational level.

TABLE 10

GENERAL ADJUSTMENT RATING ACCORDING TO AMOUNT OF EDUCATION

General Adjustment Rating	Men			Women		
	College Graduates (N = 536) %	College 1 to 4 years (N = 127) %	No College (N = 100) %	College Graduates (N = 401) %	College 1 to 4 years (N = 99) %	No College (N = 101) %
1. Satisfactory .	68.9	68.5	73.3	69.2	57.8	57.6
2. Some maladjustment ..	22.8	20.7	22.1	24.1	26.7	30.6
3. Serious maladjustment ..	8.3	10.8	4.6	6.7	15.5	11.8

In spite of the lack of relationship of schooling to general adjustment for men and the tendency for the women with less education to be less well adjusted, there is a significant positive relationship between scores on our difficult test of intelligence (Concept Mastery test) and general adjustment rating. This is true even though Concept Mastery test scores tend to increase with amount of education (see Chapter V). The difference in mean score between those rated satisfactory and those with some or serious difficulty in adjustment is significant for both sexes ($P = .001$ for men and .01 for women). Table 11 gives the mean Concept Mastery score according to the three general adjustment categories for the 551 men and 453 women who took this test in 1950–52.

TABLE 11

MEAN CONCEPT MASTERY TEST SCORE ACCORDING TO
GENERAL ADJUSTMENT RATING

General Adjustment Rating	Men			Women		
	N	Mean Score	S.D.	N	Mean Score	S.D.
1. Satisfactory	391	136.4	26.2	303	130.8	27.7
2. Some maladjustment	120	145.6	26.1	117	138.1	26.4
3. Serious maladjustment ..	40	152.8	23.8	33	140.0	29.6

SUMMARY

When the data on general adjustment are reviewed, we find a small but fairly consistent sex difference in the direction of more maladjustment among gifted women than among the men of our group. These sex differences must be interpreted with caution, especially where comparisons are made with the generality. The relatively small number of gifted subjects involved magnifies the importance of each case in comparing the incidence of various problems in our group with that reported for the total population. With these qualifications, a discussion of the sex differences follows.

A slightly larger proportion of women than men have suffered a mental disorder serious enough to require hospitalization (3.4% of women and 3.1% of men). In the incidence of suicide, although the pattern agrees with that for the total population in which more men than women commit suicide, the sex difference in rate is considerably less for the gifted than that found for the general population. In the country as a whole, the suicide rate for males is three to four times that for females; among the gifted, the rate, proportionately, is less than twice as high for men as for women.

The problem of small numbers confronts us in comparing the sex difference in excessive use of alcohol. The 10 men and 3 women who could be considered alcoholics in 1950, according to the World Health Organization definition, constitute 1.3 percent of men and 0.5 percent of women. These figures indicate a frequency three times as great for gifted men as that for gifted women. On the other hand, authorities[20] report that in the total U.S. population alcoholism is found 6 to 7 times more frequently among men than among women. The same difficulty with numbers occurs in evaluating the extent of homosexuality, where 17 men (2.0%) and 11 women (1.7%) are known to be or to have been homosexual. In contrast to this small difference, Kinsey[22] estimates for the generality that homosexuality is from one-half to one-third less frequent among females than among males.

In crime and delinquency, on the other hand, the women have a considerably better record than do the men. None has been in prison and none in reform school, though two are known to have been arrested for vagrancy.

In the broader areas of the relationship of general adjustment to education and intelligence test scores, we find some additional sex differences. Considerably more women than men who discontinued their schooling below the college graduate level are rated as having either some or serious difficulty in adjustment. There is no sex difference, however, in the general adjustment ratings of college graduates. The data on the relation of intelligence test scores to general adjustment are not consistent. For the childhood Binet IQ, maladjustment among women is more frequent at the highest level (IQ 170 and above) but there is no difference in adjustment rating according to IQ level for men. In Concept Mastery scores, on the other hand, the sex difference is in the other direction. Although the less well-adjusted of both sexes score higher on the average than do those rated satisfactory, the differences in score according to adjustment rating are greater for men.

CHAPTER V

INTELLECTUAL STATUS AT
MID-LIFE

The measure of adult intelligence has been regarded as an essential and
fundamental part of the follow-up program in our longitudinal study
of intellectually superior children. It is as important to know the relationship between adult intelligence and early mental status as it is to
compare the adult gifted subjects with both the general population and
selected populations in intellectual ability. The first adult testing was
undertaken in the 1939–40 follow-up when the subjects averaged 29.5
years of age. A test was needed that could be administered to a group
in a brief period, that was sufficiently difficult to differentiate at a very
high level, and that would yield a statistically reliable measure of intellectual functions similar to those brought into play by the Stanford-
Binet Scale and other tests highly saturated with Spearman's "g"
(general intelligence).

The Concept Mastery Test

Because there was no suitable test available, it was decided to con-
struct one that would meet our requirements. Before deciding upon the
content of the proposed test we made a survey of the results yielded by
the leading types of intelligence tests: their reliabilities, their validities
as measures of "intellect," and their relative efficiency per unit of time.
As a result of this survey two types of tests were chosen: the synonym-
antonym test and the analogies test. A large battery of items was given
to a group of 136 university students and an item analysis of the test
results yielded 190 items which discriminated reliably between the top
and bottom half on the basis of total score of the population tested. The
1939 test in its final form was divided into two parts, each arranged
according to order of difficulty. Part I Synonym-Antonym consisted
of 120 pairs of words to which the subject responds with "same" or
"opposite," and Part II Analogies was made up of 70 statements in
which the subject chooses the fourth term of the equation from a choice

of three responses. There is no time limit; however, those for whom the test is applicable usually complete it in from thirty to forty minutes.

Both the synonym-antonym test and the analogies test are of the type commonly designated as "verbal." They are not as exclusively linguistic, however, as they appear to be. It is possible to devise verbal tests that measure not only vocabulary but a wide variety of information. In the selection of items, an effort was made to tap as many fields as possible by the use of concepts related to physical and biological science, mathematics, geography, history, logic, literature, art, religion, music, sports, et cetera. The test has been named the Concept Mastery test because it deals chiefly with abstract ideas. Abstractions are the shorthand of the higher thought process, and a subject's ability to function at the upper intellectual levels is determined largely by the number and variety of concepts at his command and on his ability to see relationships between them.

It should go without saying that neither this nor any other test of intelligence measures native ability uninfluenced by schooling and other environmental factors. Like other group intelligence tests of the verbal type, its scores are probably more influenced by such factors than are scores on the Stanford-Binet. Although no amount of educational effort can furnish the naturally dull mind with a rich store of abstract ideas, it is obvious that one's wealth of concepts must inevitably reflect in some degree the extent of his formal education, the breadth of his reading, and the cultural level of his environment. The surprising thing is that despite such influences there are subjects in our gifted group with only a high-school education who score as high on the Concept Mastery test as others who have taken graduate degrees in superior universities and are successful lawyers, doctors, and college teachers.

Another merit of the Concept Mastery test is that it measures power rather than speed. Several studies have shown that the speed of mental processes declines more in middle and later maturity than does the level of power. Chess champions, for example, may retain most of their superb playing ability to an advanced age—a type of ability that demands a high degree of constructive imagination and abstract reasoning. Verbal abilities, even more notably, tend to survive the hazards of age. Sward's professors emeriti in the sixties and seventies scored as high in a difficult vocabulary test as his matched group of young university teachers ages twenty-five to thirty-five.[32] Specialized skills often atrophy from disuse, but one's thinking all life long involves the manipulation of concepts. The ability to deal with concepts is therefore one of the *fairest* measures of intellectual power from middle age onward, provided cultural opportunities have not been unduly limited. This test has been described at length in Chapters XI and XII of *The Gifted Child Grows Up.*[36]

During the course of the 1939–40 follow-up the Concept Mastery test was administered to 954 gifted subjects and to 527 of their spouses. For comparative purposes the test was also given to six additional groups comprising 466 subjects. Although there appeared to be some drop in IQ, as estimated from the CMT scores, beyond the expected statistical regression due to errors of measurement and failure of the childhood and adult tests to measure the same functions, the gifted group at a mean age of about 30 years was, on the average, still within the 98th or 99th percentile of the generality. Furthermore, comparison of the CMT scores of the gifted subjects with those of other groups tested showed the former to rank far above the average level of ability of students in top-ranking universities.

For the retesting of the group in the 1950–52 follow-up, a second form of the Concept Mastery test* was devised. Tests of reliability and validity show the second form (Form T) to be as good as, and probably superior in many ways to, the 1939 test (later designated as Form A). For the preliminary item selection, a battery of 435 new items plus the 190 from Form A were given to 764 subjects at four educational levels. These included 214 ninth-grade students, 222 twelfth-grade students, 219 college-undergraduate students, and 109 graduate students. As in Form A, items were selected on the basis of the extent to which they discriminated between the upper and lower half on the basis of total score of the populations tested. The tryout batteries made it possible not only to select the items for Form T with confidence in their validity and degree of difficulty, but also to match Form T with Form A items for both difficulty and content. Form T differs from Form A, however, in the addition of easier items extending the scale downward, and in the elimination of some of the excess top in Form A. The surplus of top in Form A was limited to the synonym-antonym section in which the 10 or 15 most difficult items were

---

* The 1950 revision was published in 1956 by The Psychological Corporation, New York City, as the Concept Mastery Test—Form T and will be referred to as Form T in this volume. In prior publications this form is referred to as Form B.

so seldom answered that they contributed little to the test. Additional items at the lower end of the scale in both sections were needed to measure adequately the less highly selected groups with which it was desired to compare the gifted, as well as to allow for possible regression with age of the gifted subjects themselves. Ten easier synonym-antonyms and 5 easier analogies were added to Form T, and 15 of the most difficult synonym-antonyms were eliminated. The final Form T scale consists of 190 items, 115 synonym-antonyms, and 75 analogies.*

Evidence of the reliability of the Concept Mastery test was obtained by correlating Form T with Form A. Test-retest data were available for four groups who took both forms. Group 1 consisted of 108 Stanford University undergraduates, and 40 graduate students and teaching assistants at the University of California. Group 2 included 341 Air Force captains tested by the Institute of Personality Assessment and Research of the University of California. The 768 gifted subjects and the 334 spouses of the gifted who took both forms made up Groups 3 and 4. For the first two groups the interval between test and retest was from one day to one week, with the order of presentation of the two forms being alternated. In the case of the gifted subjects and their spouses the reliability coefficients were obtained from a comparison of the Form A scores of 1939–40 with the Form T scores of 1950–52 for those individuals who took the CMT at both times. The time interval between testings for the last two groups was from 11 to 12 years. As shown in Table 12, the test-retest correlations for these various groups ranged from .94 to .86. Since Form A was a slightly more difficult test it may be assumed that the reliability coefficients are somewhat

TABLE 12

CORRELATIONS BETWEEN FORMS A AND T OF THE
CONCEPT MASTERY TEST

Group	N	r	Form A Mean	Form A S.D.	Form T Mean	Form T S.D.
Undergraduate and Graduate Students and Teaching Assistants (Stanford and University of California)	148	.94	74.6	36.5	95.6	39.4
Air Force Captains	341	.86	42.8	24.5	60.2	31.3
Subjects of Gifted Study ...	768	.87	96.9	29.8	136.7	28.5
Spouses of Gifted Subjects..	334	.92	62.1	35.2	95.3	42.7

* As stated earlier, Form A also contained 190 items; however, the synonym-antonyms numbered 120 and the analogies 70.

lower than would be found in correlating two comparable forms of the Concept Mastery test.

SCORES OF THE GIFTED SUBJECTS AND THEIR SPOUSES

During the 1950–52 follow-up the Concept Mastery test* was administered by the field workers to a total of 1,004 gifted subjects and 690 spouses of the gifted. Table 13 gives the distributions of scores with means and standard deviations separately for men and women of the gifted and spouse groups.

TABLE 13

CONCEPT MASTERY, FORM T SCORES OF
GIFTED SUBJECTS AND THEIR SPOUSES

	Gifted Subjects			Spouses of Gifted Subjects		
Score Interval	Men	Women	Total	Husbands	Wives	Total
180–189.........	16	4	20	3	1	4
170–179.........	60	31	91	4	8	12
160–169.........	76	46	122	18	13	31
150–159.........	84	62	146	20	18	38
140–149.........	68	71	139	17	20	37
130–139.........	73	60	133	25	25	50
120–129.........	54	52	106	23	25	48
110–119.........	34	42	76	23	33	56
100–109.........	30	23	53	15	36	51
90–99...........	20	25	45	11	35	46
80–89...........	13	18	31	19	31	50
70–79...........	9	9	18	23	31	54
60–69...........	9	5	14	21	32	53
50–59...........	2	4	6	18	26	44
40–49...........	3	1	4	13	37	50
30–39...........	..	..	..	11	14	25
20–29...........	..	..	..	5	16	21
10–19...........	..	..	..	3	7	10
0–9.............	..	..	..	..	6	6
–10 to –1........	..	..	..	1	3	4
N	551	453	1,004	273	417	690
Mean	139.4	133.4	136.7*	102.6	90.5	95.3*
S.D.........	28.8	27.7	28.5	42.4	42.2	42.7

* Coincidentally, in spite of the sizable difference in N's, the mean CMT scores for the total gifted and spouse groups as given in this table, and the corresponding means given in Table 12 for those among the total who took both forms of the test, are identical.

One-half of the gifted subjects made a score of 141 or better; the median for men was 144 and for women, 138. About 2 percent of the gifted (16 men and 4 women) had scores in the 180–189 interval (per-

* Unless otherwise indicated, the Concept Mastery scores referred to in the following discussion are Form T scores.

fect score equals 190), and 10 subjects, a scant one per cent, scored below 60. The scores of the spouses ranged from minus 8* to 189 with half of the husbands scoring above 108 and half of the wives about 92. For both the subjects and their spouses the mean scores were slightly lower than the medians, and the standard deviations were approximately 28 for the gifted and 43 for the spouses. The sex differences in mean score for the gifted and spouse groups, though comparatively small, are statistically significant, with critical ratios of 3.4 for the gifted and 3.6 for the spouses. The spouse group, in spite of an average score about one standard deviation below that of the gifted subjects, are themselves a superior group as is shown in a comparison of the data in Tables 15, 16, and 17.

RELATIONSHIP OF CONCEPT MASTERY SCORES TO EARLY TESTS

A comparison of the Concept Mastery scores earned by the gifted subjects in 1950–52 with their childhood Binet IQ shows a progression in Concept Mastery mean scores corresponding to increase in Binet IQ level. Of the 1,004 gifted subjects who took the Concept Mastery, 703 had been selected in the original survey of 1921–22 by a Stanford-Binet test. Table 14 gives the mean score and standard deviation on the Concept Mastery for these subjects classified according to their childhood Binet IQ. Despite the attenuating effects of age when first tested and despite the additional attenuation due to the highly curtailed distribution of the childhood IQ's, there is nevertheless a positive correlation of .29 between Binet scores in childhood and Concept Mastery scores 30 years later. The differences in mean CMT scores at the various Binet IQ levels are highly significant as determined by the F ratio ($P = < .001$).

RELATIONSHIP OF CONCEPT MASTERY SCORES TO EDUCATION

There is also a positive relationship between Concept Mastery score and the level of education. An increase in score with increase in amount of education is to be expected in a test of mental ability because of the corresponding increase in degree of selection on the basis of intelligence at the higher educational levels. It should be noted also that a

---

* Below zero scores result from too many wrong guesses (score equals rights minus wrongs) and may be regarded as zero scores.

TABLE 14

CONCEPT MASTERY SCORES ACCORDING TO
CHILDHOOD STANFORD-BINET IQ

		Concept Mastery Test	
Binet IQ	N	Mean	S.D.
170 and above	48	155.8	23.1
160–169	70	146.2	26.2
150–159	200	136.5	29.0
140–149	344	131.8	28.6
135–139	41	114.2	33.3

test of the type and level of difficulty of the Concept Mastery cannot be entirely free from the influence of schooling. Table 15 gives the mean scores and standard deviations for the CMT according to the amount of education for both the gifted subjects and their spouses, and Table 16 gives the corresponding data for a group of 333 Air Force captains.*

TABLE 15

CONCEPT MASTERY SCORES ACCORDING TO EDUCATIONAL LEVEL
FOR GIFTED SUBJECTS AND THEIR SPOUSES

	Gifted Subjects			Spouses of Gifted Subjects		
Educational Level	N	Mean	S.D.	N	Mean	S.D.
Ph.D.	51	159.0	19.3	12	148.7	27.8
M.D.	35	143.6	23.2	18	123.9	27.8
LL.B.	73	149.4	20.7	20	126.5	37.5
Master's or equivalent degree	151	144.3	25.4	47	130.8	32.5
Graduate study one or more years without degree	122	143.0	26.9	50	119.7	33.8
Bachelor's degree only	263	135.7	26.6	192	105.0	38.0
All college graduates	695	140.9	26.0	339	114.6	37.8
College 1–4 years	163	128.7	29.7	164	84.6	36.9
No college	146	118.4	28.5	187	68.6	38.0

TABLE 16

CONCEPT MASTERY SCORES ACCORDING TO EDUCATIONAL LEVEL
FOR AIR FORCE CAPTAINS

Educational Level	N	Mean	S.D.
College graduation	66	73.0	36.2
College 1–4 years	131	60.5	32.3
High-school graduation	121	56.7	26.8
Less than high-school graduation ...	15	34.5	19.7

* These men constitute Group 2 of Table 12 and Group 10 of Table 17 and are described on page 59. Information on schooling was lacking for 11 of the total 344 officers tested.

Although the educational attainments of the Air Force group are considerably lower than those of the gifted subjects and their spouses, a relationship is still apparent between CMT mean scores and extent of education. For all three groups the differences in mean scores, according to educational levels, are significant $(P = < .001)$.

COMPARISON OF SCORES YIELDED BY THE CONCEPT MASTERY TEST

In order to secure normative data, the Concept Mastery test was given to a number of subjects outside the gifted study. Table 17 gives the mean scores and standard deviations for all the various groups tested including the gifted and their spouses. Brief descriptions of the subjects in Groups 3 to 10 of Table 17 follow:

Group 3—*Graduate Students* (University of California Institute of Personality Assessment and Research—IPAR). This group was composed of 80 senior medical school students and 81 advanced graduate students, all of whom were within one year of completing their degrees, for the most part the Ph.D., in the Graduate Division, University of California.

Group 4—*Electronic Engineers and Scientists* (Navy Electronics Laboratory). These subjects were tested in a study of creativity made at a Navy electronics laboratory. All were college graduates and about one-third had taken some graduate work, including several Ph.D.'s.

Group 5—*Applicants for Ford Foundation Graduate Fellowships in Behavioral Sciences.* This group was composed of 83 college seniors from 34 colleges throughout the United States.

Group 6—*Undergraduate Students* (Stanford University and University of California). About 41 percent were seniors and the remainder were juniors and sophomores.

Group 7—*Graduate Students* (University of California). The majority of these students took the CMT in connection with the counseling services at the University of California.

Group 8—*College Graduates.* This was a nonstudent group. All had at least a bachelor's degree, and 9 of the total 75 held graduate degrees. They were tested at the University of California Counseling Center where they had sought vocational counsel.

Group 9—*Applicants for Admission to the Public Health Education Curriculum* (University of California). This group of college graduates from various institutions was seeking admission to the Public Health Education curriculum, a graduate department.

Group 10—*Air Force Captains* (IPAR). An evaluation study for the Air Force of captains who were up for promotion, included the CMT. The median age of these men was 33 years. The extent of education in the group is reported in Table 16.

TABLE 17

MEDIANS, MEANS, AND STANDARD DEVIATIONS OF CONCEPT MASTERY SCORES
FOR VARIOUS GROUPS TESTED

Group	Concept Mastery, Form T			
	N	Median	Mean	S.D.
1. Subjects of Gifted Study	1,004	141	137	28
2. Spouses of Gifted Subjects	690	96	95	43
3. Graduate Students (IPAR)	161	120	118	33
4. Electronic Engineers and Scientists	95	92	94	37
5. Applicants for Ford Foundation Fellowships	83	116	118	35
6. Undergraduate Students	309	92	94	33
7. Graduate Students	125	121	119	33
8. College Graduates	75	111	112	32
9. Applicants to Public Health Education Curriculum	54	101	97	29
10. Air Force Captains	344	55	60	32

A comparison of the figures in Table 17 shows that the gifted subjects, regardless of the amount of schooling, far outdistance all the other groups in mean score. This is true even though in general there is an increase in CMT score with increase in educational level. As shown in Table 15, the mean CMT score for the gifted subjects who have taken a Ph.D. degree is 159. In contrast, Group 3 of Table 17, which is made up of advanced graduate students at the University of California, had a mean score of 118. Of the 161 men in Group 3, one-half were in the final year of medical school and the other half with only a few exceptions were within one year of completing work for a Ph.D. Separately, the mean CMT score for the 80 medical students was 114 and for the 81 Ph.D. students the mean was 122. It is even more interesting to find from a comparison of Table 17 with Table 15 that the 146 gifted subjects who did not go to college at all have exactly the same mean CMT score, i.e., 118, as the advanced graduate students in Group 3 of Table 17.

The lowest scoring among the various groups of subjects reported in Table 17 are the Air Force captains but this group, too, is selected on the basis of intellectual competence though less highly so than the other groups studied. In addition to their current military rank and the fact that all were candidates for promotion, further evidence of their superiority to the generality is found in their educational record. One-fifth of this group are college graduates and close to two-fifths attended college for one to three years. Considering their generally superior educational and vocational status, it can be assumed that the mean Concept Mastery score for these men would be considerably above that found for a random sampling of the general population. The difference of 77 points between the mean score of the Air Force men and that of the gifted subjects, of whom only 1 percent score as low as the mean of the Air Force group, warrants the assumption that the gifted subjects would excel by an even greater margin the so-far-undetermined mean for the generality.

Unfortunately, we have no direct way of measuring the degree to which the gifted subjects have maintained their superiority to the general population. We lack a random sampling for the CMT as well as data for comparing Form T scores with Stanford-Binet or similar scores, other than the data for the gifted subjects themselves. The 1939 version of the CMT, Form A, however, was given to a group of college students who were also given the 1937 Stanford-Binet and the Wechsler tests.[1] By statistical inference based on a comparison of the CMT scores of this college sample with their standing on the Stanford-Binet and the Wechsler tests, it was estimated that the 1939–40 Concept Mastery scores of the gifted subjects were, on the average, 2.5 S.D.'s above the mean of the general population.

The Maintenance of Intellectual Ability[*]

Although we are not able to determine accurately the amount of change, if any, in the intellectual status of the gifted subjects since their selection in childhood solely on the basis of ability to score within the top one percent of the generality on a standardized intelligence test, we are able to compare the status of the gifted group—and of their spouses as well—at two different testings with the Concept Mastery Test: Form A in 1939–40 and Form T in 1950–52.

In order to secure equivalent scores for Forms A and T, both forms were given, either in immediate succession or at one-week intervals, to 148 subjects. Form A was given first to half of the subjects and Form T first to the other half. This sample was composed of 108 Stanford undergraduates and 40 graduate students and faculty members at the University of California. After correction for practice effects, the transformation by "line of equivalents" was made and a conversion table set up by which Form T scores could be converted

---

[*] The data to be discussed here are from a detailed study of the changes in intellectual status with age among the gifted subjects and their spouses. This study, made by Nancy Bayley and Melita Oden, is reported in "The Maintenance of Intellectual Ability in Gifted Adults," in the *Journal of Gerontology*.[2]

into Form A equivalents. These equated scores will be referred to as T(e) scores. For this sample of 148 the correlation between the two forms is .94.

In the follow-up of 1939–40 the Concept Mastery, Form A had been given to 954 gifted subjects and 527 spouses of the gifted and in the 1950–52 follow-up Form T was given to 1,004 gifted and 690 spouses. There were 768 gifted subjects (422 men and 346 women) and 335 spouses (144 husbands and 191 wives) who took both Form A and Form T. The elapsed time between the two testings was from 11 to 12 years. The average age of the subjects at the earlier testing was 29.5 years and at the later testing it was approximately 41.5 years. The average age of the spouses was about the same—41.2 years at the 1950–52 follow-up. The test-retest correlations for the 11- to 12-year interval are all high. For both gifted men and gifted women the Form A scores correlated .88 with the Form T(e) scores. In the case of the spouses, both husbands and wives, the correlation between A and T(e) scores was .92. Such difference as there is between the correlations for the gifted and the spouses can be accounted for by the greater variability in scores of the less highly selected spouse group.

Table 18 gives the Form A and Form T(e) mean scores and standard deviations separately by sex for those gifted subjects and spouses of the gifted who took both forms of the Concept Mastery. The data show that the scores of both the gifted and the spouse groups were consistently higher at the second testing when the subjects were approximately 12 years older. The increases in score for each of the subgroups were highly significant statistically, the level of significance being better than .001 in every instance. Table 18 includes also the mean score and S.D. for the gifted men and women who were tested once only on the CMT (either Form A or Form T). A comparison of the CMT scores of the twice-tested subjects with the scores of those tested only once indicates that the twice-tested are typical of the total. The increases in mean score from Form A to Form T(e) range from 11.4 points for the wives of gifted men to 16.3 points for the gifted men. The gifted women improved their score by 15.5 points and their husbands increased 14.9 points on the average. In interpreting these increases the reader is cautioned that a line of equivalents based on 148 cases is bound to have some sampling errors and these estimates of change have to be regarded as approximations.

Although the general trend was toward an increase in score on retest, not all of the gifted showed a gain from A to T(e) scores. A few

TABLE 18

COMPARISON OF TEST-RETEST SCORES ON THE CONCEPT MASTERY TEST

Twice-tested	Form A			Form T(e)			Differ-ence A:T(e)
	N	Mean	S.D.	N	Mean	S.D.	
Gifted men	422	98.6	30.8	422	114.9	26.2	16.3
Gifted women	346	94.9	28.5	346	110.4	25.3	15.5
Husbands	143	65.3	34.5	143	80.2	39.0	14.9
Wives	191	59.8	35.1	191	71.2	39.2	11.4
Tested Once Only							
Gifted men	95*	98.2	33.8	129	117.0	30.0	18.8
Gifted women	73*	90.0	27.6	107	105.4	26.7	15.4

* 10 men and 8 women not included had died in the interval between 1940 and 1951.

subjects, about 6 percent, lost more than 5 points, but except for two cases none of the losses was greater than 20 points. The man with the greatest drop (26 points) on the retest had scored within the top one percent for gifted men on Form A and in about the top 18 percent on Form T and, thus, was still well above the average of the gifted group. Since this man held a Ph.D. degree and was a college instructor, it seems reasonable to suppose that the lower score on Form T was due to some accidental circumstance at the time of testing rather than to a true change in intellectual ability. The woman with the greatest loss from Form A to Form T(e) had scored 20 points (nearly two-thirds of an S.D.) below the mean for the gifted on Form A and was more than 2 S.D.'s below the mean on Form T. She had one year of college work and a brief business course. Married and the mother of two children, she has been employed for many years in clerical work. She has few interests outside her home and job and reports that she does very little reading other than the newspaper ("usually") and popular magazines. There was little evidence of cultural interests in the home, and the subject, though cooperative in the gifted study, says she feels that she does not really belong in the group.

An investigation of the relationship between gains or losses in score and such factors as age, initial score, education, and occupation indicated that improvement occurs to about the same extent at all ages, at all levels of ability tested, and in all educational and occupational levels represented. The data from the retests of the gifted group and of their spouses (also intellectually superior on the average though less highly selected than the gifted) give strong evidence that intelligence of the type tested by the Concept Mastery test continues to increase at least through 50 years of age.

CHAPTER VI

THE MATTER OF SCHOOLING

The educational histories of the gifted subjects were reported at length in the preceding volume[86] of this series, which covered the record to 1945. Since all but a small number had completed their schooling at that time, the data presented here will (1) bring up to date the amount of schooling, (2) summarize the academic records and major fields of study which were reported in detail in the earlier volume, and (3) discuss some of the reflections on their education made by subjects themselves in the Supplementary Biographical Data blank of 1951–52.

Amount of Schooling*

The educational record of the subjects is a remarkable one: 87 percent of the men and 83 percent of the women entered college and 70 percent of men and 67 percent of women graduated. Of those who did not go to college, more than one-third of the men and one-fourth of the women had supplemented their high-school education with courses at trade, business, technical, art, or other specialized schools. On the other hand, approximately 8 percent of men and 12 percent of women did not go beyond high school and a small number of these (2 men and 9 women) did not complete the full high-school course or its equivalent. However, in these latter cases, the failure to complete high school was chiefly a formality. The two men left high school at the end of the third year, one to study music and one to enter trade school. Three of the 9 women who did not graduate from high school were child performers on the stage or in motion pictures and discontinued their schooling to study dramatics or dancing. Another three took a business school course after three years of high school. Table 19 gives the extent of schooling to college graduation, by sex, for the total group of subjects.

---

* The figures on amount of schooling, both undergraduate and graduate, given in the text and in Tables 19 and 20 are for the total group of subjects for whom the information was available. The data are given not only for those living but also for the deceased subjects who had completed their education before death. The other comparisons in this chapter and elsewhere between amount of schooling and other variables are limited to subjects living at the time the particular data under discussion were collected.

There has been a slight increase in the number of both bachelor's and graduate degrees in the 10-year period between 1945 and 1955, accounted for chiefly by the completion of their studies by those who were students at the earlier date. There were, however, a few instances in which a subject who had dropped out of college returned later to take a degree. The proportion of both sexes with no formal schooling beyond high school has remained about the same since 1940, although a few individuals in this category have continued their studies. Of particular interest is the case of one man who got his high-school diploma at the age of 45. He had left high school in 1931 while in his second year because of lack of interest and financial need. He eventually worked into a very good civil-service position for which he was able to qualify on promotional examinations even though he lacked the requisite high-school diploma. In 1955 he was awarded a high-school diploma, earned through attendance at evening classes in the adult education program, and he is now planning university extension courses to qualify himself for a position in a more technical or scientific field.

The record of nearly 70 percent college graduation in our group is especially outstanding when we consider that this was achieved chiefly in the decade 1930–40 when less than 8 percent of the generality of comparable age were graduating from college. Even in these later years when the number of college graduates has increased so markedly, the record set by the gifted subjects is far above the 12 percent of the present-day youth who complete college.[43] Especially noteworthy is the number of gifted women who are college graduates. As shown in Table 19, the proportion of women graduates is almost as great as the proportion of men graduates, namely, 67 percent of women as compared with 70 percent of men. Among the generality of college graduates, according to Wolfle's estimate (1953), the sex ratio is 60 men to 40 women.

The educational record of the gifted subjects is even more impressive when we consider the number who have continued for graduate study. Table 20 shows the extent of graduate training completed and the degrees taken. Two-thirds of the men and almost three-fifths of the women who completed college entered graduate school, and 56 percent of the men and 33 percent of the women took one or more advanced degrees. It should be noted that more have taken master's and professional degrees than show in Table 20, since each person appears only at his highest degree. The highest degree was most often the

TABLE 19

AMOUNT OF UNDERGRADUATE SCHOOLING

	Men N	Men %	Women N	Women %
College graduation	579	70.0	429	66.7
From 1 to 4 years of college (no degree) ...	139	16.8	105	16.3
High school plus special training	38	4.6	28	4.4
High school graduation	69	8.4	72	11.2
High school not completed	2	0.2	9	1.4
Total	827		643	
Deceased and education not completed at time of death	19		11	
Subjects "lost" and information lacking	11		17	
Total number of subjects in study	857		671	

LL.B., taken by 90 men; next in order of frequency was the master's (87), followed by the Ph.D. (80). The M.D. degree was taken by 49 men, and other professional degrees or diplomas by 17 men. In the case of women the master's degree held by 90 is the most frequent graduate degree. A Ph.D. degree has been taken by 17 women, the M.D. by six, and the LL.B. by two. Other graduate degrees or diplomas are held by 28 women. Slightly more than one-tenth of the men and one-fourth of the women who entered graduate school did not take a graduate degree. In some cases the advanced study was begun with the idea of getting a master's or Ph.D. degree but the students were deterred from completion by such circumstances as change of interests, lack of finances or, more fortuitously, an attractive job opening. Often, especially among the women, the graduate work was taken to qualify for a professional credential, usually in teaching, rather than for a degree.

Interesting comparative data on graduate study and graduate degrees for the college population in general are furnished by Wolfle[43] who reports a steady increase since 1900 in the number of United States college graduates who enter graduate school. By 1953, approximately one-fourth of those who received bachelor's degrees proceeded to graduate study. Of these, about 17 percent took a master's degree and about two percent a Ph.D. or comparable doctorate. On the other hand, in the gifted group, nearly all of whom had completed their undergraduate study by 1940, two-thirds of men and about three-fifths of women with a bachelor's degree entered graduate school and, as the 1955 record

TABLE 20

PROPORTION OF COLLEGE GRADUATES WHO PROCEED TO GRADUATE STUDY
AND DEGREES TAKEN

	Men		Women	
	N	%	N	%
Ph.D. or other doctorate	80	13.8	17	4.0
M.D., M.Sc.D.	49	8.5	6	1.4
LL.B., LL.M.	90	15.5	2	.5
Master's plus 1 or more years of study	26	4.5	20	4.7
Master's degree	61	10.5	70	16.3
Other graduate degrees or diplomas	17	2.9	28	6.5
Graduate study for one or more years and no degree taken	63	10.9	107	24.9
One or more years of graduate work.....	386	66.7	250	58.3
Bachelor's degree only	193	33.3	179	41.7
Total college graduates	579		429	

shows, 56 percent of the men and 33 percent of the women have
taken one or more graduate degrees. The most striking contrast is
seen in the number of doctorates: 14 percent of gifted men and 4 percent of gifted women (10% for the sexes combined) as compared with
the current 2 percent of the generality of college graduates (men and
women) who get a doctor's degree.

Nor did education end for the gifted when they left school. No
mention has been made in either Table 19 or Table 20 of the large number who have participated in adult education or workshop programs
or who are otherwise engaged in special study on a noncredit basis. In
addition, postgraduate certificates, licenses, and diplomas have been
received by a number of subjects who qualified by examination for such
special designations as Fellow, American Board of Surgery; Diploma,
American Board of Examiners in Professional Psychology; Fellow,
American Institute of Architects; Certified Public Accountant; and
similar certifications. Also special study in arts, crafts, foreign languages, literature, etc., are frequently mentioned among the avocational
interests of our subjects.

COLLEGE RECORDS AND MAJOR FIELDS OF STUDY

In general, the academic record for the college graduates was superior, with 78 percent of men and 83 percent of women having an average
grade of B or better in college. One or more honors (graduation *cum
laude*, Phi Beta Kappa, or Sigma Xi) were won by 40 percent of men

and 33 percent of women graduates. On the other hand, 53 men and 10 women flunked out of college. Of these, 24 men re-entered and graduated and 15 continued for advanced degrees, including 4 Ph.D.'s, 3 law degrees, 2 medical degrees, and 1 engineering degree. Only one woman of the 10 who failed later completed college, and she graduated *cum laude*. No clear pattern of causes for college failure emerges; however, the most common explanation given by the subjects was that in high school they had found it so easy to make high marks that they underestimated the amount of study necessary in college. A few felt that they had missed the fun and social life of high school, and on entering college devoted too much time to extracurricular student body or social activities. There were also a good many instances in which the poor record could be attributed to lack of proper guidance in the selection of a major field.

For men the five leading major fields were as follows in order of frequency: Social Sciences, 42 percent; Physical Sciences, 17 percent; Engineering, 15 percent; Biological Sciences, 10 percent; and Letters, 9 percent. In the case of women, although the Social Sciences led with 37 percent, it was by a slim margin. Letters was in second place with 36 percent. Next in order for the women were Education, 9 percent; Biological Sciences, 6 percent; and Physical Sciences, 5 percent. No more than 2 percent of either sex majored in any of the other fields of study reported.

The most frequent major among men who took graduate work was Law, with 25 percent of the total. Physical Sciences (13%) and Engineering (8%) combined were in second place with 21 percent. Next in order of choice were Social Sciences, 18 percent, followed by Biological Sciences, with 17 percent. Education and Letters each claimed 6 percent. There were no more than 3 percent in any other graduate major field. For women graduate students the five leading major fields were the same as in the undergraduate years but the order changed. The graduate majors and the percentage in each are as follows: Social Sciences, 32 percent; Letters, 28 percent; Education, 19 percent; Biological Sciences, 9 percent; and Physical Sciences, 5 percent.

EDUCATION IN RETROSPECT

The Supplementary Biographical Data blanks which the gifted men and women filled out in 1951–52 offer some interesting sidelights on the educational history of our group. In reviewing their education at mid-life, the great majority reported themselves as satisfied with their schooling. When asked to identify in a list of 10 the factors that had contributed to their life accomplishment, adequate education was the factor most frequently checked by both sexes (83% of men and 79% of women). In reply to the question: *"Did you have as much schooling as you wanted? If not, explain,"* 71 percent of men and 62 percent of women gave an unqualified "Yes" and an additional 6 percent of men and 10 percent of women replied with a qualified "Yes" such as "Yes, at the time, but not now," or "Yes, but the wrong kind." A small number (1.8% of men and 0.6% of women) stated only that their schooling was of the "wrong kind." The inability to finance further schooling was the most frequent explanation of a "No" response (13% of men and 16% of women). Some 7 percent of both men and women cut short their schooling because of the war, marriage, lack of interest, or lack of encouragement. Finally, reasons of health were given by a small minority (2.3% of men and 3.5% of women).

When we compare the replies to this question with the actual amount of schooling, we find, as would be expected, that the great majority of college graduates (84% of men and 88% of women) had as much education as they wanted. Some interesting sex differences appear among those who did not attend college at all. Almost half of the men who did not enter college felt that they had had as much education as they wanted, while only 32 percent of women with no college attendance were satisfied with their education. Table 21 gives the distribution of replies for three levels of education.

TABLE 21

Subject's Feeling About Amount of Schooling
According to Educational Attainment

	MEN			WOMEN		
Did you have as much schooling as you wanted?	College Graduate (N = 456) %	One to 4 Years College (N = 80) %	No College (N = 64) %	College Graduate (N = 325) %	One to 4 Years College (N = 77) %	No College (N = 78) %
Yes (unqualified) ...	80.7	45.0	28.1	80.6	29.9	16.6
Yes (qualified)	3.7	5.0	18.8	7.1	19.5	15.4
Wrong kind	1.8	3.8	....	0.6	....	1.3
No, lack of finances..	6.6	26.2	39.1	5.5	33.8	44.9
No, lack of encouragement or motivation	2.2	5.0	9.4	1.5	1.3	11.5
No, miscellaneous reasons	5.0	15.0	4.6	4.7	15.6	10.3

According to the retrospective opinions of the subjects in the biographical data blank, the attitudes of their parents had been favorable to educational achievement. The subjects were asked to report on their parents' attitude toward (*a*) school progress, (*b*) school work, (*c*) college attendance. Their replies are given in Table 22. In only a few cases did the parents fail to give encouragement or to show interest in the school work of these children. That the parents of 88 percent of men and 85 percent of women should have encouraged college attendance for their sons and daughters in the 1920's and 1930's, when so small a proportion of their classmates were continuing beyond high school, is truly remarkable. The fact that more than twice as large a proportion of women as of men say that they were not encouraged to go to college because of financial difficulties can be explained on the ground that girls could not be expected to find ways to earn their expenses as readily as boys, and because of the parental view that money for a college education was not as well spent for girls as for boys. On the other hand, more than three times as many men as women said that their parents were indifferent, or left the decision to the subject, or that college attendance was not considered in the family. More than half of the men who reported this passive attitude on the part of their parents, however, *did*

TABLE 22

PARENTS' ATTITUDE TOWARD SCHOOLING OF THE GIFTED SUBJECTS

Attitude of Parents	Men (N = 601) %	Women (N = 483) %
(*a*) Toward school progress		
1. Encouraged to forge ahead	51.0	46.2
2. Allowed to go own pace	47.7	52.4
3. Held back	1.3	1.4
(*b*) Toward school work		
1. Demanded high marks	12.1	13.8
2. Encouraged or expected high marks	84.6	81.7
3. Little concern or interest	3.3	4.5
(*c*) Toward college attendance		
1. Encouraged college	87.7	85.3
2. Indifferent; decision left to subject	3.7	1.0
3. College not encouraged because of subject's youth, health, considered waste of money, etc.	1.2	1.7
4. College not encouraged because of limited finances	4.3	9.7
5. College not encouraged—subject did not explain	3.2	2.3

graduate from college, and almost one-fourth continued for graduate study.

As one might expect, the great majority of those who say that they were encouraged to go to college by their parents did so: 84 percent of the men and 78 percent of the women in this category took a bachelor's degree, and 56 percent of the men and 44 percent of the women continued for graduate work. Among the subjects whose parents did not encourage college, we find that 10 percent of men and 14 percent of women were graduated. Table 23 gives the amount of education according to the parents' attitude toward college attendance.

TABLE 23

AMOUNT OF EDUCATION ACCORDING TO PARENTS' ATTITUDE
TOWARD COLLEGE ATTENDANCE

	Encouraged		Indifferent		Not Encouraged	
	Men (N = 527) %	Women (N = 412) %	Men (N = 22) %	Women (N = 5) %	Men (N = 52) %	Women (N = 66) %
College graduation	83.5	77.6	50.0	(3 cases)	9.6	13.6
College 1 to 4 years.....	11.6	15.6	27.3	......	23.1	18.2
No College	4.9	6.8	22.7	(2 cases)	67.3	68.2

It must be remembered that the feelings and opinions expressed in the biographical data blanks are memory reports. There is no way of knowing the part that rationalization may have played or the extent to which these opinions may have been colored by later experiences.

Another favorable circumstance that no doubt contributed to the high rate of college attendance in our group was the superior educational attainments of the parents themselves. Approximately 35 percent of fathers and 16 percent of the mothers held a bachelor's degree or better, while 11 percent of fathers and 15 percent of mothers had attended college for from one to three years, often receiving a certificate or professional degree. Since 88 percent of our subjects were born before 1915 and 38 percent before 1910, the schooling of the majority of the parents took place in the 1890's and early 1900's. In contrast, Wolfle reports that in 1900 in the country as a whole only 1.7 percent of persons of college age graduated from college.

CONCLUSION

From the foregoing it is evident that the educational attainments of the gifted subjects are not only far above those of their contemporaries but also well ahead of the present-day rates in terms of college degrees, both undergraduate and graduate. But, good as these educational records were, they could have been better. The figures given are more significant when they are read in reverse. All of these subjects were potentially superior college material yet more than 10 percent of the men and more than 15 percent of the women did not enter college, and 30 percent of the group did not graduate. Inability to finance college and lack of parental encouragement were frequent causes of failure to attend college, but more often the real cause was the failure of the high school to recognize the gifted students' potentialities and to give the needed encouragement and stimulation.[37] This research has not yielded a great deal of information on methods and techniques for the education of the gifted since the investigation was not undertaken as a study in the pedagogy of gifted children, but rather as a search for the basic facts necessary to a full understanding of the gifted individual and his potentialities. Not until the physical and mental characteristics and the developmental tendencies of intellectually superior children have been definitely established is it possible to plan intelligently for their education. Only a very few of the children in this group had enjoyed any special educational opportunities in the elementary or secondary schools beyond the opportunity to skip an occasional grade. As a result of this skipping the group as a whole was somewhat accelerated, with about one-half of the boys and three-fifths of the girls graduating from high school before the age of 17. In fact, nearly one-fourth of the group completed high school by age sixteen and one-half years. A careful study of acceleration in the group was made in a search for factors that might be associated with rate of school progress. Specifically, the accelerated (defined as high-school graduation before age sixteen and one-half) were compared with the nonaccelerates on a number of variables both in childhood and in later life. These results were discussed in full in Volume IV of this series.[36] In general, the findings favored the rapidly promoted children. It was concluded that children of IQ 135 or higher should be promoted sufficiently to permit college entrance by the age of seventeen at latest, and that a majority in this group would be better off to enter at sixteen. Acceleration to this extent is especially desirable for those who plan to complete two or more years of graduate study in preparation for a professional career.

CHAPTER VII

THE MATTER OF CAREER

The extent to which the gifted subjects in their adult careers have fulfilled the promise of their superior intellectual endowments and educational attainments is one of the most significant aspects of this study. Now, at mid-life, at an average age of 44, the subjects can be considered to be pretty well established in their life work. While promotions and advances in both position and income may be expected for some years, there is not likely to be much change in field of work. This is true at least for the men. The occupational future of the women is less predictable. Some women now employed may decide that they would prefer more home and family life while some housewives may decide on taking jobs as their children grow older. Because the career patterns and the types of occupations of men and women in the gifted group are so different, as are those of men and women in general, their vocational careers will be discussed separately.

Occupational Status of Men

The occupations of the employed men were classified according to the Minnesota Occupational Scale,[16] the same scale as that used in classifying the occupations of the gifted men in 1940 and 1950. This scale contains a list of about 350 occupations which are grouped into seven categories ranging from the professions to unskilled labor. We are concerned here with Groups I to V only since no gifted men fall in either Group VI (slightly skilled) or Group VII (unskilled). Group I is limited to the professions, strictly defined to include only vocations calling for a high degree of specialized knowledge, training, or creativity. Group IV includes all agricultural and related occupations regardless of the scale of operations. The remaining three occupation groups— II, III, and V—cover occupations related to business, finance, and industry as well as the arts and entertainment, and the semiprofessional, protective service, and other nonprofessional and nonfarming occupations.

73

Group II is made up of the managerial ranks in business and industry, the officials in public or private administration, and the semiprofessional occupations. Also in Group II are the men classified as "business professional." These are men trained in the professions such as engineers, statisticians, accountants, and the like, who are working in the field of business or industry. Group III includes the owners and managers of smaller businesses, clerical and sales workers, skilled workers, both craftsmen and foremen, as well as certain service workers. From Group III we move downward to Group V which comprises the minor clerical or minor business occupations and the semiskilled trades. Table 24 gives the occupational status of the men as reported in 1955.

TABLE 24

OCCUPATIONAL STATUS OF GIFTED MEN

A. Classification according to Minnesota Occupational Scale

		N	% of Classified
Group I	Professional	345	45.6
Group II	Managerial, official, and semiprofessional....	308	40.7
Group III	Retail business, clerical, skilled trades, and kindred	83	10.9
Group IV	Agriculture and related occupations	12	1.6
Group V	Semiskilled occupations	9	1.2
	All Groups	757	

B. Not employed, or less than full-time employment

	N	% of Total
1. Incapacitated by reasons of health	9	1.1
2. Independent means, or retired	9	1.1
3. Temporarily not employed	2	0.3

C. 1955 occupation not ascertained

(includes 11 men lost since 1928 or earlier).............	18	2.3
Total	795	

More than 86 percent of the employed men are in the professions of Group I and the higher business and semiprofessional occupations of Group II, and only 11 percent are in Group III. Groups IV and V are very much in the minority with 1.6 and 1.2 percent, respectively. Table 25 gives the breakdown of the occupational groups and indicates the percentage of employed men in each subgroup.

The most frequent profession is law, with 10 percent of all gifted men either practicing law or in judicial positions. An additional 9 men with the LL.B. degree were admitted to law practice but have since gone into other occupations. Next most frequent among the professions are members of university faculties and these are closely followed by engineers. Although the proportion of practicing physicians in Table 25 is 5 percent, it should be noted that an additional 6 men with the M.D. degree are full-time members of university faculties and are classified in that category. The number of clergymen also is not fully represented in the 0.9 percent reported in Table 25 since two former ministers have joined college faculties, one at a theological college and one at a secular university. Although these men may occasionally serve as ministers they are primarily teachers. The 11 men in the "other professions" include three clinical psychologists, two dentists, two landscape architects, two foresters, one biologist, and one librarian.

The largest among the subgroups by a slight margin (1 man) are the executives in business and industry of Group II, with 79 men as compared to the 78 practicing lawyers in Group I. The addition of the banking, finance, and insurance executives brings the business executive representation to better than 16 percent of the employed. These two leading Group II occupations are composed of men having executive and administrative responsibilities in broad areas of management and on policy-making levels. With the exception of the two higher business occupations and the top four professions, no other subgroup includes as many as 5 percent of the employed. As shown in the list of occupations in Table 25, the vocational interests of the gifted men have led them into many fields and many kinds of work.

TABLE 25

BREAKDOWN OF OCCUPATIONAL GROUPS

		N	%
I.	Professional occupations		
	1. Lawyers (include judges)	78	10.3
	2. Members of college or university faculties...........	57	7.5
	3. Engineers	55	7.3
	4. Physicians (practicing)	40	5.3
	5. School administrator or teacher (high school or junior college).................	32	4.2
	6. Chemists and physicists	27	3.6
	7. Authors or journalists	17	2.3
	8. Architects	8	1.1
	9. Geologists and kindred	7	.9
	10. Clergymen	7	.9
	11. Economists	6	.8
	12. Other professions	11	1.5

TABLE 25—(*Continued*)

	N	%
II. Managerial, official and semiprofessional		
1. Higher business		
Executives and managers in business and in industry	79	10.4
Executives in banking, finance, and insurance........	44	5.8
Accounting, statistics, market research, *et cetera*....	35	4.6
Sales (sales managers, technical, or engineering sales)	24	3.2
Advertising, publicity, public relations.............	19	2.5
Personnel, labor relations, vocational placement, and kindred	10	1.3
Building and construction (owners or officials)......	9	1.2
Office manager, department head, *et cetera*.........	12	1.5
2. Arts and entertainment		
Radio, television, or motion pictures: producer, director, writer	21	2.8
Musician or actor	6	.8
Applied arts (illustrator, commercial art, decorator, and kindred)	6	.8
3. Semiprofessions		
Draftsman, surveyor, and kindred	9	1.2
Nonacademic teaching (trade school, technical, avocational, *et cetera*)	4	.5
Other	3	.4
4. Army and Navy officers	15	2.0
5. Officials in administration (public or private)		
Includes government officials	12	1.6
III. Retail business (small), clerical and sales, skilled crafts, protective services (supervisory ranks) and kindred		
1. Clerical and sales	29	3.8
2. Skilled crafts (craftsmen and foremen)............	29	3.8
3. Retail business (small) owners and managers.......	10	1.3
4. Technicians (laboratory assistants, dental technicians, *et cetera*	5	.7
5. Protective service occupations*	10	1.3
IV. Agriculture and related occupations		
1. Farm owners and operators......................	11	1.5
2. Nurseryman	1	.1
V. Minor business and semiskilled occupations		
1. Clerical and business	7	.9
2. Semiskilled trades	2	.3
Total	757	

* Includes 1 Inspector and 1 Captain of Police; 3 Police Sergeants; 1 Sheriff; 2 Battalion Chiefs (fire department); 1 Chief Warrant Officer, U.S. Army; and 1 Master Sergeant, U.S. Army.

The following case notes will illustrate some of the occupations that are classified in the various groups. The descriptions do not cover all occupations represented in these groups but we have attempted to show something of the variety in both the nature and the level of occupation.

Group I

(1) A director of engine research for one of the largest companies manufacturing heavy equipment. His story is especially interesting because he arrived at his present high-level engineering job without the usual training. He graduated from high school with honors at age 17 and immediately went to work as an office boy for the firm with which he is still associated. He was made an apprentice mechanic within a few months and thereafter rose rapidly through the various engineering levels to his present post.

(2) A newspaper reporter and columnist who has also written a number of books on sports for children.

(3) A physician and specialist in cancer research who heads the department of internal medicine in a leading university.

(4) A pastor of a church in a medium-sized city who has taken a doctorate in theology. He is engaged in pastoral work and heads the district council of churches.

(5) A man with a B.S. in physics who is employed as a physicist in the research and development division of a large oil company.

(6) A university professor of astronomy engaged in teaching and research. Under various grants, including a Guggenheim Fellowship, he has carried out research projects in Africa as well as at several observatories in this country and has published a number of research papers.

(7) Two writers of interest because of different backgrounds and parallel careers. One trained in engineering (B.S. and M.S.) is one of the country's leading science fiction and fantasy writers who has produced some 60 short stories and novelettes as well as 15 volumes of fiction and nonfiction. He is also the author of a number of magazine articles, critiques and book reviews dealing chiefly with science or science fiction.

The other, with a background in the liberal arts (A.B. and M.A.), is both prolific and versatile. By age 40 he had produced 7 mystery novels, 25 or more short stories, and 10 or 12 articles. A reviewer of note, he writes a regular column of book reviews for a nationally circulated metropolitan newspaper. He has won the Edgar Allen Poe award four times for the best mystery story reviewing in the United States. He is the editor of three magazines of mystery or science fiction and has also published five anthologies of mystery stories. In addition, he has taught a class in creative writing for the past seven years.

(8) A former high-school teacher who is now supervisor of measurement and evaluation in a large city school department.

(9) A member of the editorial staff of a leading journal who was formerly a professor of fine arts and more recently associate director of an Institute of Fine Arts.

Group II

(1) A man who took a Ph.D. in physics with high scholastic honors and then found himself more interested in economic theory and political science. After several years experience in statistical economics with private firms, he was in government service during World War II in charge of industrial control programs. After the war he joined a world-wide shipping company where he has now become controller and vice-president.

(2) A motion picture director who has made some of the most outstanding pictures of the past ten years. His pictures made in England and on the Continent as well as those made in the United States have won him an international reputation. He has received a number of citations and awards including a special award at the International Film Festival. Other honors include several "Oscar" awards from the Academy of Motion Picture Arts and Sciences won either by his pictures or by actors under his direction.

(3) A graduate in chemistry (B.S.) who joined the sales division of a large chemical company. After several years in the field of technical sales work, he received a year of additional training in economics from the company and is now in charge of a sales division.

(4) A graduate in engineering (A.B. and M.E.) who went into the construction business. He has been highly successful in the field of residential property development.

(5) A former high-school teacher of mathematics and science who is now a land surveyor.

(6) A city manager of a small California city. This man majored in political science and did graduate work in personnel administration. He formerly worked in the field of merchandising.

(7) A public relations official with one of the branches of the state government who is a writer by avocation and whose first novel was a recent Book-of-the-Month selection.

(8) The director of the textbook division for a large publishing firm.

(9) The chief of police of a small city. He has developed a co-ordinated communications system, which is considered a model. A special interest in the prevention of juvenile delinquency has brought him frequent invitations to lecture on this and other aspects of his work.

Group III

(1) A university graduate and former botanist who operates a small, independent pest-control business.

(2) An art-school graduate who paints in both water colors and oils and whose work has received considerable praise in exhibits, is employed

as a house painter and decorator while hoping some day to win recognition as an artist.

(3) A man with a B.S. in chemistry who is a photographic technician.

(4) A high-school graduate who is a supervising clerk in a public utilities office.

(5) A high-school graduate who is a pattern-maker and instructor of apprentices.

(6) A high-school graduate who is a battalion chief in a large city fire department.

(7) A noncommissioned officer in military service who is also a short story writer specializing in science fiction. Eight of his stories have been published to date.

Group IV

The 12 men in this category with the exception of one employed as a nurseryman are all owners or operators of farms or ranches. Six are orchardists, three are cattle ranchers, one is a poultry farmer, and one a rice- and grain-grower. Eight of these men are college graduates, two attended college for two years, and two are high-school graduates.

Group V

(1) A university graduate who also had two years of graduate work, who is now a mail carrier. Because of poor health this man is not able to work at his profession.

(2) Two high-school graduates, both of whom are bartenders and managers.

(3) Two men with college educations who are in minor clerical work. In both cases lack of stability and excessive drinking have prevented greater vocational achievement.

(4) Two men who did not complete high school, one of whom is a truck driver and one a warehouseman.

(5) A high-school graduate who operates a small sandwich shop.

(6) The ninth man in Group V is described on page 82.

COMPARISON WITH OCCUPATIONS OF MEN COLLEGE GRADUATES IN GENERAL

There is no question but that the gifted men have many times the representation in the professional and higher business occupations than would be found for a random group of men of like age. That this group also surpasses in occupational status unselected college graduates is shown by a comparison with the data furnished by Havemann and

West.[17] The authors have grouped "professionals of all types" into a single category, which presumably includes semiprofessionals and business professionals as well as the more strictly defined professions of the Minnesota Scale. Since this is too broad a grouping to be comparable to the Group I in which professional gifted men are classified, we have combined the two Havemann and West categories, "professionals of all types" and "proprietors, managers, and executives," for comparison with the combined Groups I and II of the gifted men as classified on the Minnesota Scale. Table 26 gives the percentage distribution of occupations for three groups of men: the U.S. college graduates in general, the gifted college graduates and, in addition, all gifted men regardless of education.

TABLE 26

COMPARISON OF OCCUPATIONS OF GIFTED MEN
AND U.S. COLLEGE GRADUATES

Occupational Classification	All U.S. College Graduates %	Gifted College Graduates %	All Gifted Men %
A. Professionals of all types; proprietors, managers, executives	84	93.7	86.3
B. Clerical, sales, and kindred workers	10	4.3	10.9
C. Skilled, semiskilled, and unskilled workers	5	0.5	1.2
D. Farmers and farm workers	1	1.5	1.6

A comparison of the figures in Table 26 shows the superiority in job status of the gifted graduate to college men in general. This is evident in the higher proportion of the gifted in the important and high-level occupations (94% of gifted versus 84% of all college men) and in the very much smaller proportion of gifted graduates among both the clerical and sales workers and the skilled, semiskilled, and unskilled group. Not only do the college graduates among the gifted surpass the generality of college graduates in occupational status, but the total group of gifted men, including the 30 percent who did not graduate from college, also compare favorably with Havemann and West's college graduate population. As shown in the third column of Table 26, the gifted men, regardless of education, have a slightly larger representation in the higher occupations of category A and a markedly smaller proportion of workers at the "skilled, semiskilled, and unskilled" level of category C where 5 percent of unselected college graduates fall as compared with one-half of one percent of gifted college men and 1.2 percent of all gifted men. The proportion in farm occupations shows little difference in the three groups of men.

OCCUPATIONAL CHANGES AMONG GIFTED MEN
BETWEEN 1940 AND 1955

A comparison of the occupational status of the gifted men at different stages illustrates their upward progress. The fact that the Minnesota Occupational Scale was used in classifying the occupations of 1940, 1950, and 1955 makes it possible to compare the status of the men at these three dates.* These comparisons are given in Table 27.

TABLE 27
COMPARISON OF OCCUPATIONAL CLASSIFICATIONS
OF 1940, 1950, AND 1955 (MEN)

	Percent of Employed Men†		
Occupational Group	1940 (N = 724)	1950 (N = 762)	1955 (N = 757)
I. Professional	45.4	45.6	45.6
II. Managerial, official and semiprofessional	25.7	39.4	40.7
III. Retail business, clerical, skilled crafts and kindred	20.7	12.0	10.9
IV. Agriculture and related occupations	1.2	1.8	1.6
V. Semiskilled occupations	6.2	1.2	1.2
VI. Slightly skilled trades	0.7	...	...

† The slight variations in the number employed at each period are caused, in part, by the student status of some subjects at the earlier dates; in part, by deaths during the 15-year span; and in a very few instances by incapacitation due to poor health at one time or another.

Omitting Group IV (farming and related occupations), the rank order of Groups I, II, III, and V in Table 27 remains unchanged throughout the 15-year period. There are, however, marked changes in the proportionate representation in all groups except the professions with most of the change taking place between 1940 and 1950. Although the proportion in Group I is about 45 percent at all three dates, there have been some shifts into and out of the professions, which are not apparent in the percentages. The majority of those who were graduate students in 1940 subsequently entered the professions, and a few men formerly in other occupational classifications took additional training following World War II to qualify for one of the professions. On the

* The classification made of the 1944 occupations as given in the 1945 Information Blank is omitted from these comparisons because of the occupational dislocations caused by the war.

other hand, about an equal number of men who were in Group I in 1940 have left the active practice of their profession. All but a few of these men have become staff or operational executives in business or industry. They are classified according to the current occupation, chiefly Group II, although their work may be related to the former profession and require an application of their professional training. There are, however, five men who were in professional work in 1940 who are currently in Group III. These include an artist and writer who has turned to carpentry to earn a livelihood; two school teachers, one of whom, because of special skill with tools, became a machinist, and one who is manager of a small business; an engineer who operates a small radio and television repair service; and a lawyer who, for health reasons, is doing clerical work.

In the 15-year interval from 1940 to 1955, the Group II representation increased from 26 percent to 41 percent; Group III dropped from 21 percent to 11 percent; Group V decreased from about 6 percent to approximately 1 percent, and Group VI disappeared entirely. Only nine men are currently in Group V and the number in this category is not likely to be reduced greatly since, in all probability, there will always be a few individuals who, for personal reasons, choose more simple and routine work. In spite of the marked increase in Group II between 1940 and 1955, four men formerly in Group II have moved to a lower classification. Three of these are now in Group III, all in smaller retail business enterprises. The fourth man shifted to Group V and his story is especially interesting.

A university graduate, D. H. for some 15 years worked for a large corporation in which he rose to a managerial position. His growing interest in the labor movement and in bettering the social order made the atmosphere of "big business" in which he worked increasingly uncongenial. Finally, after World War II he left his position to go to work as a laborer and has continued in this work for the past 10 years. He is very much interested in history, economics, and politics especially as applied to the labor union and the working man, and reads extensively in these areas. D. H. takes an active part in union affairs and appears to have found great personal satisfaction in what he is doing. His score on the Terman Group Test at age 15 was equivalent to an IQ of approximately 150 and this high intellectual level has been maintained. On the Concept Mastery tests taken in 1940 and 1952 he scored at both dates well above the average of the gifted men.

It is interesting to see what happened to the men who in 1940 were classified in Groups III, V, and VI and this information is given in Table 28. Of the total 200 men in these three occupational groups in 1940, 10 had died by 1955, and information on 1955 occupation was not obtained for 4 men. There were no drops in the occupational level among the remainder of the men in the 15-year period; instead, for all but a minority there was marked improvement. Almost four-fifths of the men classified in either Groups III, V, or VI in 1940 moved upward and one-fifth remained in the same classification between 1940 and 1955. About two-thirds of those in Group III at the earlier date had advanced to the higher business and professional occupations (Groups I and II) and one-third of the 1940 Group V men were classified in Groups I and II in 1955.

TABLE 28

1955 OCCUPATIONAL STATUS OF MEN WHO WERE CLASSIFIED
IN GROUPS III, V, AND VI IN 1940

1955 Occupational Status	1940 Occupational Classification		
	Group III N	Group V N	Group VI N
Group I	14	5	..
Group II	84	10	1
Group III	37	26	2
Group V	..	1	2
Not employed; independent means...	1	..	..
Incapacitated by reasons of health....	3	..	..
Information on status lacking	4	..	..
Deceased	7	3	..
Total in group as of 1940.........	150	45	5

It is not surprising, of course, that the gifted men who have had the advantage of college training, often at the graduate level, should be in positions of importance and prestige in the professions and the business world. It is, however, of special interest when those without such educational advantages rise to positions of importance in competition with college-trained men. One such example follows.

C. J., whose formal schooling was limited to high school and 6 units of college mathematics taken in extension courses, moved from Group III to Group I between 1940 and 1955. He was one of a family of two children (brother and sister), both of whom were selected for the gifted study. For various reasons the boy, although he had had a strong interest in science and engineering since childhood, did not go to college. The fact that he completed high school at a time of economic stress (the early 1930's) may have been one determining factor behind his failure to enter college. His parents, although they had hoped he would continue his education, were not able to help him financially. More important, however, were his poor school grades, which made it necessary that he take "make-up" courses to qualify for college, and at that time he could see no reason for spending time on subjects in which he was not interested. Probably the most crucial factor in his dropping out of school was the failure of the school itself to recognize his unusual ability or to offer any real guidance during his high-school years. His reluctance to conform to a school routine and his lack of application to his studies—even though, according to the report from the high school, he showed "occasional flashes of brilliance"—apparently obscured his great gifts. Left entirely on his own with little sympathetic stimulation and no guidance, he went to work on leaving high school, with the intention of saving money for college study and a degree in engineering. It was an unfavorable time for financial progress, but C. J. remained employed all through the depression. He began at a fairly unskilled level but after a few years found work in the field of machine design where he made excellent progress. During this period he studied informally and still clung to his ambition of taking an engineering degree and—as he came to hope—a graduate degree in physics. When his income became sufficiently secure that he might have gone to college, war threatened and he turned instead to war work. During World War II he was on the research staff of a highly secret and important laboratory, working side by side with graduate physicists, often on his own projects, an honor usually accorded only Ph.D.'s. When this research laboratory was discontinued at the end of the war, he was appointed to the engineering staff at a military ordnance laboratory. Because of his fine work as a project engineer on important military developments, he received a promotional appointment to the GS-12 level under a "meritorious exception." This was a distinct honor since, under Civil Service regulations, an individual without a college degree is ineligible for advancement beyond the grade of GS-7.

However, greater honors were in store for C. J. He was recently fully qualified as a mechanical engineer, GS-12, thus removing the "meritorious exception" qualification. This action made further promotion possible and he now heads a branch of the optical engineering division in a military research and development center. His work, on a high professional level, is concerned with guided missile instrumentation.

C. J. is now in his early forties, married, and the father of three children. He is active in school and community affairs and his hobbies include music, photography, and reading. Among the magazines read regularly are the *Atlantic Monthly* and *Scientific American*, and books he has recently read include *Modern Arms and Free Men* (Vannevar Bush), *Language in Action* (Hayakawa), and *Human Destiny* (Lecomte du Noüy). His Binet IQ at age 10 was 154 and his Terman Group Test score in 1928 at age 17 was within a few points of a perfect score. And it was then that the school complained of his argumentativeness and failure to respond to discipline, and noted his failure in various school subjects, despite the A he received in chemistry! On the Concept Mastery tests taken in 1940 and 1951 he scored far higher than the average college graduate and placed nearly 20 points above the average of the gifted men. In view of his continued high intelligence rating and his remarkable scientific ability—especially in physics and engineering—one wonders how much farther he might have gone and how much greater might have been his contribution to knowledge had his talents been recognized early and adequate guidance and motivation been provided.

Occupational Status of Women

According to the 1955 reports, one-half of the gifted women were housewives with no outside employment, 42 percent held full-time jobs, and about 8 percent were working part-time. The occupational picture changes, however, when viewed according to marital status. Only 29 percent of the married women were working wives on a full-time basis and 10 percent had part-time employment. The three single women not holding regular jobs are financially independent and engage in volunteer welfare work and various creative activities, such as painting, dress design, and writing. One of the women in this category has published several books for children; another, for several years was a designer for an exclusive dress shop. Four-fifths of the divorced and widowed were employed full-time and 1 percent part-time. Table 29 gives the occupational status according to marital status.

TABLE 29

OCCUPATIONAL STATUS OF WOMEN ACCORDING TO
MARITAL STATUS (1955)

	Marital Status							
	Single		Married		Widowed or divorced		Total	
	N	%	N	%	N	%	N	%
Housewife, not employed....	..	..	290	61.2	13	18.1	303	49.7
Employed full-time.........	59	92.2	138	29.1	56	77.8	253	41.5
Employed part-time........	..	..	46	9.7	1	1.4	47	7.7
Independent means, not employed	3	4.7	..	..	..	..	3	.5
Incapacitated by health reasons	2	3.1	..	..	2	2.7	4	.6
N	64		474		72		610	

Status not ascertained (includes 17 women lost since 1928 or earlier).. 19

Total 629

Occupational status is associated to some extent with the amount of education as shown in Table 30. The women who have taken advanced degrees are much more likely to be employed than are those with only a bachelor's degree or those who had from one to four years of college work. All but two of the 25 women with a Ph.D., M.D., or LL.B. were following careers either on college faculties or in professional practice. One woman with an M.D. and one with an LL.B., both of whom previously engaged in professional practice, are now housewives. For the total college graduate group the proportion employed is 43 percent as compared with 37 percent of the nongraduates. It might be noted here that although only 29 percent of the currently married women are employed, 79 percent of them are college graduates.

TABLE 30

OCCUPATIONAL STATUS OF WOMEN ACCORDING TO EDUCATION

	Percentages in Educational Categories				
	Graduate degree (N = 137)	Graduate study 1 or more years, no degree (N = 104)	Bachelor's degree (N = 175)	1 to 4 years of college (N = 100)	No college (N = 94)
Housewife, not employed	36.5	40.4	61.7	57.0	48.9
Employed full-time	59.1	43.3	30.8	32.0	43.6
Employed part-time	4.4	15.4	6.3	7.0	7.5
Single, not employed....	..	0.9	0.6	1.0	..
Incapacitated	..	..	0.6	3.0	..

The employed women represent a wide range of occupations that do not fit into a formal classification such as the Minnesota Occupational Scale used for the gifted men. Their occupations have been grouped instead into three broad categories: (1) professional and semi-professional; (II) business occupations; and (III) a small group of miscellaneous occupations not covered in either Group I or Group II. Table 31 lists the 1955 occupations and gives the percentage of employed women in each. The 17 women on college faculties include, in addition to 11 who are in the fields of the humanities and social sciences, four biologists, one zoologist, and one biochemist. Five are full professors and 3 of these five also hold administrative positions: one as Dean of the Faculty and Provost, one as Dean of Women, and one as chairman of her department. Another 5 women hold associate professor rank and 4 are assistant professors; the remaining 3 faculty members are research scientists, all in the biological sciences. Grouped together as the "higher professions" are those vocations which require specialized training and graduate degrees. These include six physicians, three clinical psychologists, two lawyers, and a research metallurgist. The category "other professions" in Group I includes two pharmacists, two laboratory technicians, a missionary, an aerodynamist, and a court reporter. Group III is composed of occupations that do not fit into either of the other groups. These include telephone operator, sales clerk, and industrial worker.

Schoolteaching, including elementary and secondary administrative and supervisory positions, is the most frequent occupation, accounting for almost one-fourth of the employed women. In second place are the secretaries, stenographers, and similar office workers with about one-fifth of the total. The women on college faculties or in the higher professions rank third with 11 percent. The remainder are distributed over a number of occupation, all with less than 10 percent representation.

TABLE 31

OCCUPATIONS OF WOMEN WITH FULL-TIME EMPLOYMENT

I. *Professional*	N	%
Members of college or university faculties..............	17	6.7
Higher professions	12	4.7
Administrators and supervisors, elementary or high school	7	2.8
School teachers or counselers	53	20.9
Social workers	20	7.9

TABLE 31—(*Continued*)

Authors or journalists..............................	17	6.7
Librarians	14	5.5
Arts and music	7	2.8
Nurses	6	2.4
Economists, statisticians, and kindred................	5	2.0
Other professions	7	2.8
II. *Business*		
Secretary, stenographer, bookkeeper, or other office workers	50	19.8
Executive and managerial...........................	20	7.9
Public relations, advertising, promotion...............	6	2.4
Real estate and investments	5	2.0
III. *Miscellaneous*	7	2.8
Total employed	253	

Although fewer than one-half of the women were engaged in careers outside the home in 1955, and the jobs they chose have often been of a less demanding type in order not to interfere too greatly with home-making, there have nevertheless been some remarkable vocational accomplishments. A few examples of outstanding careers follow.

Several scientists have made important contributions to research to the extent that 7 women are listed in *American Men of Science*.[5] Five of these are in the field of the biological sciences and two in the social sciences. Among the distinguished biological scientists is one who played an important part in the development of the vaccine for polio-myelitis and who is continuing to work in virus research. The social and behavioral sciences include several women who are outstanding in the fields of psychology, education, and social welfare. Two of the women psychologists are among those listed in *American Men of Science*.

Only one woman is working as a high-level physical scientist. Her undergraduate major was engineering and her graduate work was taken in a related physical science. Since taking her Doctor of Science degree, she has worked in private industry where she has successfully competed with men for advancement and is now one of the most highly paid women in our group. In addition to research and the publication of many technical papers, she has taken out three patents. This woman is also a gifted linguist and has translated several of her own more important works into French and German.

But not all the conspicuous achievements have been in the sciences. Several outstanding records have been made in the humanities and arts as well as in the business world. One of our most distinguished women is a gifted poet whose work has received wide recognition and who is rated among the outstanding poets of our day. Her writing has appeared in a number of literary magazines as well as in several anthologies. She has published four volumes of poetry and various essays, critiques, and monographs in the field of literature and philosophy. Others among the women writers are a feature article writer who contributes to leading magazines, two novelists, a member of the editorial staff and an executive editor of a nationally circulated magazine, and still another is the editor of a small literary magazine. Other writers include the author of a successful Broadway play (also produced as a motion picture), several journalists including a reporter and feature writer for a metropolitan daily paper, and several technical writers. One of our women is a gifted painter whose work has appeared by invitation in many exhibits and who has won considerable recognition. Several women have been phenomenally successful in business; two of these are in the real estate business, another is an executive buyer in a large department store, and still another, herself a pharmacist, is the owner and operator of a prosperous pharmacy.

One of the most interesting, versatile, and successful of the women has had three careers in addition to that of housewife and mother. This subject, F. B., was first tested for the gifted group at the age of 7 years 10 months and was found to have an IQ of 188. At that early age she was already showing remarkable literary ability. She had begun making up stories and rhymes at the age of three but few of these early compositions were preserved. At six years she was given a typewriter and began recording her own work. From the age of 6 to 11 she frequently illustrated her stories and poems by crayon or water-color drawings. Her childhood poetry was compared most favorably by a board of judges with literary juvenilia by Tennyson, Blake, Longfellow, Wordsworth, Shelley, and others. Until the age of 11 she read and studied at home under her mother's direction. She then entered the ninth grade of a private girls' school and completed high school at 14½ years. High-school graduation was followed by entrance at a coeducational university where she was graduated three years later at the age of 17. During her high-school and college years her interests turned to clay modeling and painting, both oil and water color. However,

writing continued to be her main interest and her first novel (not published) was completed before her eighteenth birthday.

After college F. B.'s interests were directed more and more to art,
particularly sculpture, although she did take time to write a short novel,
which was published in 1938. In the mid-1930's she went abroad and
spent several years studying sculpture with an outstanding master.
During this period she produced several pieces of sculpture which have
received high praise. She was only 27 years old when the threat of
World War II forced her to return to the United States, where she continued her work in the arts, not only in sculpture but also in design and
decoration. F. B. was married during the war and accompanied her
husband to the military posts at which he was stationed. After her son
was born and the family was again settled, she became interested in
real estate investments. She began buying, remodeling and redecorat-
ing older houses and then selling to advantage. She developed this
interest into a highly successful financial venture, which she continued
for some years. Lately, she has given more time to her artistic work
and recently completed, as a volunteer service, a mural in sculpture for
an interracial youth center. This work has received much attention
and praise not only for its artistic merit, but also for its originality in
the use of novel materials and techniques, hitherto untried.

The many interests of this woman—her literary career, her later
business career, and her continuous work in the arts—have not interfered with her fourth career as a housewife. She has maintained a
home, directed her son's development, and shared in her husband's
interests, all successfully.

There have been many others among the women whose achievements, though important, have not received the public recognition of
the examples given above. Among these are two missionaries, both of
whom have visited us and given first-hand accounts of their work in
foreign lands. Each determined at an early age to become a missionary
and planned her education accordingly. S. B. was the valedictorian of
her high-school graduating class and was also voted the best all-round
girl in the school on the basis of the following characteristics: the most
intellectual, the wittiest, the friendliest, and the best natured. After
college graduation she taught school for a few years before being sent
to China as a missionary. Her first years in this work were spent in a
country village and from there she joined the staff of a teachers college. In all, S. B. spent 10 years in China, the last two and a half under

the Communist regime. When eventually the mission was withdrawn owing to the unfavorable political situation, she returned to the United States. After a year of graduate work at the university she was appointed to a teaching position with the Institute of International Education in Indonesia, but it is still her hope to return some day to China.

W. H., after receiving a teacher's certificate, joined the Sudan Interior Mission and was sent to Nigeria where she has worked for the past 20 years except for occasional furloughs. After some years teaching in village schools, she organized a teacher-training school for natives. Her most recent post has been supervisor of schools in northern Nigeria. These are, for the most part, one-teacher schools in widely scattered bush areas. Hers has been a life of tremendous work and many hardships, but a rich and rewarding one. One of her first undertakings was to learn the native language and she has translated many religious works and written several books in this language. Her writings include in addition to a number of religious tracts and devotional booklets, a translation of the story of Dr. Carver, a book on the life of St. Paul, and a book on the history of the missions.

The distinguished records and notable accomplishments, however, have not been limited to the so-called "career" women. We should not overlook the women whose careers have been limited to the role of wife and mother or to community welfare and civic betterment on a volunteer basis. Among the former is H. M., one of our most highly intelligent women. Her childhood IQ of 192 was one of the highest in the entire group and later tests also placed her at or near the top of the group. This subject, especially talented in mathematics, took her A.B. in astronomy with honors at the age of 20 and continued her studies for a master's degree in the same field. This was followed by marriage to a fellow scientist at the age of 22. She taught science in a junior college for two years after her marriage until the birth of her first set of twins. In a period of 11 years she and her husband became the parents of eight children, including three sets of twins. This has left little time for any kind of professional career or even community activity, although she has maintained her interest in science and reads widely in both scientific and more general areas of literature. With two children still in the toddler stage, her only activity outside the home has been P.T.A. In view of this subject's extraordinarily high IQ in childhood, her Concept Mastery scores are of especial interest. Because this family has lived in the east since 1940, it was not possible to give Form A until 1948 and Form T was given in 1950, both during visits to Stanford University. At both testings H. M. made scores close to the highest of any made by the gifted subjects, and her husband, who took only Form A, made an almost equally high score.

A number of housewives, less occupied with children than the woman just described, have found time to contribute to community and civic activities, many as leaders, holding various positions of responsibility. Among these is J. M. who, by the age of 37, not only had been elected president of the national alumnae association of her college, but also had been elected to the college Board of Trustees.

This subject was married at the age of 23 to another member of the gifted group who has become one of the most successful lawyers in our group. For the first few years after her marriage, J. M. worked as a private secretary but for the past ten years her time has been given to volunteer work with such organizations as the American Cancer Society, the Girl Scouts, and the Crippled Children's Society. Since she has no children, she has been able to give considerable time to service activities as well as to such organizations as the World Affairs Council and the League of Women Voters. She has been especially active in the American Cancer Society, holding important executive office and board memberships at both state and local levels.

She has also published a book (a historical biography) and has served as editor of organization bulletins and as publicity writer for the groups with which she works.

The foregoing examples are not typical of the gifted women; few, if any, are so versatile in talent as F. B.; no one else, scientist or not, has eight children, and while there may be others as active in service and community work as J. M., none has combined this with such honors as election to a college Board of Trustees and the publication of a book. However, most of the women in our group, both housewives and career women, have engaged in many activities and have followed a variety of interests outside their homes and jobs, which are described at greater length elsewhere.

INCOME OF THE GIFTED SUBJECTS

Although money is certainly not the only or final criterion of success, it cannot be denied that financial reward is our best objective

measure of progress on the vocational ladder. The information schedules at each follow-up investigation since 1936 have asked for a report
of earned income of both the subject and the spouse, and in recent years
the amount of income from sources other than earnings has also been
called for. The income data were requested not only for the calendar
year immediately preceding the time of inquiry but also for the years
that had elapsed since the last follow-up, thus giving us a continuous
record. The significance of any given income is determined by the
economic climate in which it is received; this is especially true during
periods of mounting inflation such as have prevailed during the last
decade. By current standards incomes that stood high in relation to
the average a few years ago may now seem mediocre or even low. For
this reason our discussion of income, both earned and total family income, will be limited in this chapter to the most recent figures available;
that is, the 1954 annual income as reported in the 1955 questionnaire.

Earned Income of Men

Adequate information on earned income was supplied by 673 of the
men, and Table 32 gives the median and percentage distribution of
earnings both for the total group and for subgroupings according to
age. Four-fifths of the group were 40 to 49 years old with more than
half (57%) under 45 years of age and 10 percent under age 40. By age,
the highest median income ($10,283) was reported by the 45- to 49-
year-old men, and, as was to be expected, the lowest income was that of
the 30- to 39-year age group. For the total group, ages combined, the
1954 median earned income was $9,640 and the income ranged from
around $4,000 to $400,000 per year. As shown in Table 32, 10 percent
of men (all ages) earned $25,000 or more in 1954. The incomes of the
67 men who constituted the top 10 percent were distributed cumu-
latively as follows by income level:

Earnings	N	%	Earnings	N	%
$100,000 or more	5	0.7	$50,000 or more	13	1.9
$ 75,000 or more	6	0.9	$25,000 or more	67	10.0

One of the men in the $75,000 to $100,000 bracket whose 1954 income was $95,000 had, in each of the three preceding years, exceeded
$100,000; in fact, the six persons with highest incomes, ranging from

$400,000 to $95,000, had each averaged $100,000 or more per year for the three-year period 1952–54.

TABLE 32

1954 EARNED INCOME BY AGE (MEN)

Earned Income	Percent of Total (N = 673)	Percent at each age who earn specified amounts			
		30–39 years* (N = 71)	40–44 years (N = 313)	45–49 years (N = 228)	50–54 years (N = 61)
$25,000 and over.....	10.0	7.0	9.9	11.8	6.6
15,000–24,999.......	17.5	14.1	18.8	16.2	19.7
10,000–14,999.......	20.2	16.9	19.5	23.3	16.4
9,000–9,999........	6.4	4.2	8.0	4.8	6.6
8,000–8,999........	8.0	4.2	7.7	9.7	8.2
7,000–7,999........	9.7	15.5	9.6	8.8	6.6
6,000–6,999........	11.9	11.3	12.8	10.1	14.7
5,000–5,999........	7.9	14.1	6.7	7.4	8.2
Less than $5,000....	8.5	12.7	7.0	7.9	13.1
Median earned income	$9,640	$7,773	$9,780	$10,283	$8,900

* Includes 5 men age 30–34 and 66 men age 35–39.

The relationship of earnings to educational attainment is shown in Table 33. The number of M.D. and LL.B. degrees will not agree with the number of men in those professions in 1954 since some of the physicians are full-time members of medical-school faculties and some of the lawyers have left the practice of law for other fields. For the same reasons, the median incomes by degree for these two professions will not agree with the median reported for men actually practicing these professions in 1954. In considering the role of education in income it is interesting to find that those who did not go beyond high school have the same median income as those college graduates who had one or more years of graduate study beyond the bachelor's but did not take a graduate degree. Furthermore, those men who had 3 to 4 years of college work but who did not graduate, exceeded both groups just mentioned. In fact, only one of the six men ranking highest in income was a college graduate. The top man ($400,000) did not attend college at all, another had 1 year of college, and three had between 2 and 3 years of college work. But these are very exceptional cases. The median income for all college graduates is 37 percent greater than that of those who did not complete college; namely, $10,725 for the college graduates as compared with $7,812 for the nongraduates.

TABLE 33

1954 EARNED INCOME BY AMOUNT OF EDUCATION (MEN)

	N	Median	\$15,000 or more	\$10,000–14,999	\$7,000–9,999	\$5,000–6,999	Less than \$5,000
			Percent earning specified amounts				
Ph.D.	74	\$ 8,917	21.6	18.9	37.8	14.9	6.8
M.D.	42	22,000	76.2	11.9	9.5	2.4	...
LL.B.	74	15,250	51.4	24.3	12.2	9.5	2.7
M.B.A.	18	11,430	22.2	38.9	16.7	16.7	5.6
Master's or other professional degree	75	8,572	18.9	18.9	32.4	17.6	12.2
Graduate study without graduate degree	45	8,167	17.8	22.2	24.4	22.2	13.3
Bachelor's degree	162	9,850	27.0	22.0	24.5	20.1	6.3
Total college graduates	490	10,725	31.8	21.2	24.4	15.8	6.8
3–4 years college	67	8,333	15.1	18.2	31.8	27.3	7.6
1–2 years college	34	7,500	17.6	17.6	17.6	29.4	17.6
High-school and special training ...	30	6,750	6.7	16.7	23.3	33.3	20.0
High-school graduation or equivalent..	52	8,167	20.8	16.7	22.9	29.2	10.4

Table 34 lists in rank order according to median income the occupations in which 5 or more men are engaged. The figures illustrate the considerable variation in the amount to be earned in particular occupations even among men as highly selected both in mental ability and in education as our group. For example, among the university faculty group, which ranks 17th in income with a median of \$8,167, the top man with earnings of \$25,000 is only slightly above the average for the physician group. The physicians in professional practice (full-time medical school faculty not included) rank first with a median annual net income* in 1954 of $23,500, followed by executives in major business or industry with $17,680, and the producers, directors, and writers in the entertainment field are next with $17,500. Lawyers with $15,970 and architects with $15,000 are in fourth and fifth places, respectively. At the bottom of the list are the clergymen with a median annual income of $4,500. Although practicing physicians rank first in *median* income, the highest individual incomes are to be found in several other fields. The 6 men in approximately the top one percent of the total group whose 1954 earnings ranged from just under $100,000 to $400,000 include two men who were in the field of land development and home building, including insurance and financing; a motion picture director, a television writer and producer, a playwright, and a business executive engaged in manufacturing and the development of oil properties.

TABLE 34

RANK ORDER OF OCCUPATIONS ACCORDING TO 1954
EARNED INCOME (MEN)

(Includes only fields in which 5 or more men were engaged)

	Occupation	N	Median
1.	Physicians (practicing)	36	$23,500
2.	Executives in major business or industry	73	17,680
3.	Radio, TV, or motion picture arts: producer, director, engineer, writer, et cetera	19	17,500
4.	Lawyers	65	15,970
5.	Architects	8	15,000
6.	Economists	5	13,750
7.	Executives in banking, real estate, finance, insurance	41	12,500
8.	Owners and executives in building and construction trades	9	11,500
9.	Chemists and physicists	26	10,835
10.	Musicians and actors (i.e., performers)..	5	10,830
11.	Geologists and related.................	7	9,750
12.	Personnel, labor relations, vocational placement officials	9	9,500
13.	Advertising, publicity, public relations..	17	9,250
14.	Engineers	45	9,100
15.	Sales managers, technical or engineering salesmen	22	9,000
16.	Army or Navy officers	14	9,000
17.	College or university faculty...........	52	8,167

* Net income of the self-employed includes earnings after deductions for business expenses but before income taxes.

TABLE 34—(*Continued*)

Occupation	N	Median
18. Authors or journalists...............	14	8,000
19. Accountants, statisticians, and kindred occupations	26	7,600
20. Office managers, purchasing agents, traffic managers, et cetera............	10	7,500
21. School teachers or administrators......	29	6,792
22. Draftsmen, surveyors, specification writers, et cetera	7	6,250
23. Owners and managers, retail business...	11	6,125
24. Protective service occupations*	11	5,900
25. Agricultural occupations	9	5,850
26. Skilled trades, craftsmen, and foremen..	25	5,700
27. Clerical and retail sales occupations.....	25	5,125
28. Clergymen	5	4,500

* Members of police and fire departments with rank of sergeant and above and noncommissioned officers in military services.

We do not have adequate comparative data on income. What is needed is a breakdown of earned income of the male population in general by age, education, and occupation, for 1954. Although we lack the information for a precise comparison, the figures available indicate that the gifted men are doing well financially. The *Statistical Abstract of the United States*[41] gives the 1954 median income for occupational categories grouped according to the occupation of the head of the spending unit. The figures are as follows:

a. Professional and semiprofessional $6,020
b. Managerial 5,800
c. Self-employed 5,710
d. Clerical and sales 3,980
e. Skilled and semiskilled 4,390

Categories *a, b, and c*, although perhaps more inclusive, can be considered roughly comparable to the gifted occupational groups I and II where the median 1954 *earned* income was $10,556. In contrast to the medians of around $4,000 for categories *d* and *e*, the gifted men in corresponding occupations (Groups III and V of the Minnesota Occupational Scale) had a 1954 median earned income of $5,750. Although the gifted men have not reached the age of peak earnings in the professional and business occupations, in the above comparisons they are at an advantage since their group does not include the younger or older workers. It must be remembered that these figures are only approximations, but the margin of difference is wide enough to permit the con-

clusion that the gifted men are above average in earnings, compared
with the generality of men in like fields.

EARNED INCOME OF WOMEN

Information on 1954 earned income was received from 184 of the
253 fully employed women. Table 35 gives the median income and the
percent distribution according to income level. The average income of
the employed women was $4,875. Earnings of $10,000 or more were
reported by 6 percent of the women and an income of $5,000 or more
was reported by almost half (47%) of the group.

TABLE 35

1954 EARNED INCOME BY INCOME LEVEL FOR FULLY EMPLOYED WOMEN

Earnings	N	%
$10,000 or more..........	11	6.0
7,000 or more..........	31	16.8
6,000 or more..........	50	27.2
5,000 or more..........	86	46.7
4,000 or more..........	134	72.8
3,000 or more..........	162	88.0
Less than 3,000	22	12.0

Median earned income............ $4,875

A comparison of income by amount of education shows that the
median for college graduates is $5,217 as compared with $3,666 for
those with one to four years of college and $4,250 for those who didn't
go to college at all. It is interesting also to find that the women who took
only a bachelor's degree (no graduate work) were earning slightly less
in 1954 than the high-school graduate group, namely, $4,056 as com-
pared with $4,250. Higher degrees, however, are accompanied by
higher incomes and the total college graduate group earns substantially
more than either of the nongraduate groups. Annual incomes (1954) of
$6,000 or more were reported by more than two-thirds of those with a
Ph.D., M.D., or LL.B. and by about 38 percent of the women with a
master's or similar degree, and by a third of all college graduates. Earnings of the gifted women according to amount of education are given
in Table 36. Education appears to be a less important factor in earnings in the case of the gifted women than among women in general.
Havemann and West[17] report that women college graduates earn more
than two and one-half times as much as the average working woman.
In the case of the gifted, however, those women with a college degree

earned only about one-fourth more than did the gifted women who did not go beyond high school.

TABLE 36

1954 Earned Income by Amount of Education (Women)

	N	Median Earned Income	Proportion Earning Specified Amounts			
			$6,000 or more		Less than $4,000	
			N	%	N	%
Ph.D., M.D., LL.B..........	19	$7,500	13	68.4	0	....
Master's or professional degree	48	5,455	18	37.5	5	10.4
One or more years graduate study without graduate degree (may include teaching credential or similar certificate)	32	5,333	10	31.3	4	12.5
Bachelor's degree only......	39	4,056	5	12.8	19	48.7
All college graduates.......	138	5,217	46	33.3	28	20.3
1 to 4 years college (not graduated)	18	3,666	1	5.6	12	66.7
No college	28	4,250	3	10.7	12	42.9

The relationship of income to occupation is shown in Table 37. The women on college faculties and those in the higher professions (law, medicine, and so on) averaging $6,833 have the highest income. For the 15 faculty women in this group the median is $5,700. Teachers (below college) and school administrators are in second place with a median of $5,400, followed by miscellaneous professions with $4,710 and, ranking last, the business occupations with a median of $3,719. The highest individual income ($24,000) is that of a physician. Others earning $15,000 or more are: another physician, a research scientist (industry), a pharmacist (owner), a lawyer, and a hospital administrator.

TABLE 37

1954 Earned Income by Occupation (Women)

(Includes only fields in which 15 or more women were engaged)

	N	Median
College faculties and higher professions..	25	$6,833
School teachers or administrators.......	48	5,400
Other professions	51	4,710
Business occupations	60	3,719

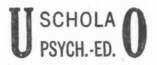

TOTAL FAMILY INCOME

In addition to data on the earned income of the subjects, the 1955 Information Blank also called for the earned income of the spouse and for the approximate amount of family income from sources other than earnings (investments, trust funds, *et cetera*). The three sources of income, i.e., earnings of subject, earnings of spouse, and income from other sources, were combined to arrive at the total family income. Total family income includes not only the earnings of the fully employed but also those of part-time workers. Single persons are not included in the presentation of data on family income.

Table 38 gives the median income and the percent distribution by income levels of the 1954 family incomes. Median total family incomes of $11,582 and $9,740 were reported by the men and women, respectively. For all families, i.e., gifted men and women combined, the median was $10,866. Nearly one-third had incomes of $15,000 or more and 12 percent were in the $25,000-and-over bracket.

TABLE 38

1954 TOTAL FAMILY INCOME

	Gifted Men and Spouses (N = 644) %	Gifted Women and Spouses (N = 454) %	All Gifted Subjects and Spouses (N = 1,098) %
$25,000 or more......	13.1	9.6	11.7
15,000 or more......	34.1	23.7	29.8
10,000 or more......	61.1	48.3	55.8
7,000 or more......	84.3	73.7	80.0
5,000 or more......	96.0	90.2	93.6
Less than 5,000......	4.0	9.7	6.4
Median family income	$11,582	$9,740	$10,866

Just as the individual earned income of the gifted subjects compares favorably with that for the generality, so also does total family income. We have compared the family incomes of the gifted subjects with the incomes of the U.S. urban white families who resemble the gifted group more closely in socio-economic status than does the total population. According to census figures,[40] the median total money income in 1955 for the economically favored urban white families was $5,069. One percent had incomes of $15,000 or higher and 63 percent of families had incomes below $5,000. As shown in Table 38, the 1954 median family income for the gifted subjects was more than twice that of the white urban group ($10,866 versus $5,069). Furthermore, incomes of $15,000 or more were reported for 30 percent of the gifted families and only 6 percent of the gifted fell below $5,000 in family income. Again, it should be pointed out that age is not controlled in these comparisons.

More About Life Work

To supplement the information already presented about the occupations of the gifted men and women and their vocational achievements and success as measured by honors, recognition, and financial reward, further information about vocations in terms of personal satisfactions has been provided in the Supplementary Biographical Data Blank. This blank, filled out during the 1950–52 follow-up, included the following question: *Which of the following best describes your feeling about your present vocation?* (Check) Deep satisfaction and interest....; Fairly content....; No serious discontent, but do not find it particularly interesting or satisfying....; Discontented but will probably stick it out....; Strongly dislike and hope to change..... Additional space was provided for comments or elaborations of response.

Approximately half of the men expressed deep satisfaction and interest in their work and another 37 percent stated they were fairly content. Only a small minority (6%) reported themselves seriously discontented. Even more women than men were content with their vocational choice; over 55 percent found deep satisfaction in their vocations and only 3 percent reported serious discontent. Table 39 gives the distribution of responses to this question separately for 600 men and 428 women.

When the opinions on vocational satisfaction are examined accord-

TABLE 39
Subject's Feeling About Vocation

Feeling about Vocation	Percentages	
	Men (N = 600)	Women (N = 428)
1. Deep satisfaction and interest.......	49.2	55.4
2. Fairly content	37.2	35.2
3. No serious discontent but not particularly satisfying	7.6	5.9
4. Discontented but will probably stick it out	5.0	2.6
5. Strongly dislike and hope to change	1.0	0.9

TABLE 40

Relationship of Vocational Satisfaction to Occupation (Men)

Feeling about Vocation	Lawyer	Executive in higher business or industry	Engineer	Physician	Member of college faculty	Accountant, statistician, and allied	School teaching or administration	Clerical or sales work	Chemist or physicist	Banking, finance, insurance	Skilled trades	Advertising, promotion, public relations	Author or journalist
						Percentages in Selected Occupations							
1. Deep satisfaction and interest..	50	47	38	84	57	47	47	28	50	57	27	41	40
2. Fairly content	41	43	45	11	34	25	37	41	50	43	41	59	33
3. No serious discontent but not particularly satisfying	6	8	12	..	3	22	7	14	..	..	14	..	7
4. Discontented but will probably stick it out	3	2	5	5	6	3	3	17	..	..	18	..	13
5. Strongly dislike and hope to change	..	..	..	..	..	3	7	..	..	..	..	..	7
Number of respondents.........	66	49	40	37	35	32	30	29	28	28	22	17	15

ing to particular occupations, some interesting differences appear. The comparisons for men have been limited to those occupations in which replies were received from 15 or more persons. Table 40 gives the percentage distribution of vocational satisfaction according to occupation for men, and Table 41 shows the relationship of vocational satisfaction to occupational status for women.

As shown in Table 40, those men who are in such Group III occupations as the skilled crafts, or clerical and sales work have the smallest proportion (28% and 27%, respectively) who report deep satisfaction and interest in their work. An additional 41 percent of men in each of these occupations, however, said they were fairly content and none expressed strong dislike or desire to change. Of all occupations, the physicians had the highest percentage (84%) in category 1 (deep satisfaction), and only 5 percent of physicians expressed vocational discontent. On the other hand, all of the men in the banking and finance occupations, all those in the advertising and public relations group, and all the chemists and physicists, fell in categories 1 or 2.

The women in the professional and semiprofessional occupations are the most satisfied group vocationally, with almost 95 percent reporting themselves at least content, and two-thirds expressing deep satisfaction in their work. Somewhat more than one-half of full-time housewives reported satisfaction in their work and 37 percent said they were fairly content. Least satisfied among the women were the part-time workers, nearly all of whom are also housewives; however, the number of persons in the part-time category is too small for the figures to be very meaningful. It should be noted that of the 74 women who omitted this item in the questionnaire, 67 were housewives.

TABLE 41

RELATIONSHIP OF VOCATIONAL SATISFACTION TO OCCUPATION (WOMEN)

			Full-time Employment	
Feeling about Vocation	Housewife, no outside employment (N = 215) %	Part-time employment (N = 23) %	Profes- sional and semipro- fessional (N = 129) %	Business and miscel- laneous (N = 61) %
1. Deep satisfaction and interest	53.1	43.5	65.9	44.2
2. Fairly content	36.6	39.1	28.7	42.6
3. No serious discontent but not particularly satisfying	6.6	13.0	3.9	6.6
4. Discontented but will probably stick it out	2.8	4.4	1.5	3.3
5. Strongly dislike and hope to change	0.9	...	...	3.3

Since only a small minority of men and women express discontent with their vocation, comparisons of vocational satisfaction with other variables are not very reliable. However, we have compared vocational satisfaction with earned income for men. (The relatively small number of women with earnings coupled with the concentration at categories 1 and 2 for vocational satisfaction do not warrant such a comparison for women.) For men we find the expected trend of decline in vocational satisfaction with decline in earned income. The figures are as follows:

	Gifted Men	
Feeling about Vocation	N	1949 Median Earned Income*
Deep satisfaction and interest	271	$7,741
Fairly content	208	6,718
No serious discontent	38	6,000
Discontented or strongly dislike	34	5,500

* The reader is reminded that the income used here is that for the year 1949 as given in the 1950 General Information Blank. The 1949 income for the gifted, just as for the generality, is considerably less than that for 1954 reported elsewhere in this chapter. Because data on vocational satisfaction were obtained in 1950–52, the 1949 income is more pertinent to these comparisons.

There is a much more marked relationship between earned income of men and the subject's opinion on how well he has lived up to his intellectual abilities. The biographical data blank asked the subjects to check in a list of six the answer that best described the extent to which they felt they had, on the whole, not just vocationally or economically, lived up to their intellectual abilities. The six response choices ranged from "Fully" to "A total failure"; however, no one among the gifted men checked the last response. The median 1949 earned incomes according to opinion on this variable are as follows:

Intellectual abilities lived up to:	N	1949 Median Earned Income
Fully	23	$11,875
Reasonably well	335	7,355
Considerably short	165	6,339
Far short or largely a failure	46	4,917

A comparison of the occupational status of women with their opinion on how well they have lived up to their intellectual abilities brings out some interesting differences. The figures given in Table 42 show that more than two-thirds of the women in the professional and semiprofessional occupations thought that they had made use of their intellectual abilities either "Reasonably well" (63%) or "Fully" (6%). Slightly more than half of the housewives and slightly less than half of the women in business and related occupations considered that they have lived up to their intellectual abilities. At the other extreme, 9 percent of the housewives and 10 percent of the business women replied that they have fallen far short or were total failures. There were also 38 women (29%) among the professional and semiprofessional group who felt that they were considerably short of living up to their intellectual abilities and three women at this occupational level thought that they had fallen far short of realizing their potentialities. The number of part-time workers is too small to be conclusive; however, they resembled the housewives in their replies to this item.

TABLE 42

OPINION ON HOW WELL INTELLECTUAL ABILITIES LIVED UP TO
BY OCCUPATION (WOMEN)

| | | | | | Full-time Employment | | | | | |
Extent abilities lived up to	Housewife N	%	Part-time employment N	%	Professional and semiprofessional N	%	Business and miscellaneous occupations N	%	Total N	%
1. Fully	7	2.7	1	3.7	8	6.1	1	1.6	17	3.5
2. Reasonably well	131	49.6	14	51.9	82	62.6	29	46.0	256	52.8
3. Considerably short	102	38.6	9	33.3	38	29.0	27	42.8	176	36.3
4. Far short ...	19	7.2	2	7.4	3	2.3	3	4.8	27	5.6
5. Consider life largely a failure	4 ⎱ 1.9		1 ⎱ 3.7				1 ⎱ 4.8		6 ⎱ 1.8	
6. A total failure	1 ⎰		⎰				2 ⎰		3 ⎰	
Total	264		27		131		63		485	

Another item in the biographical data blank asked the subjects to check in a list of ten those aspects of life from which the greatest satisfaction was derived. "Your work itself" was in first place for men with mention by close to four-fifths (78%) of those replying. Income, on the other hand, ranked sixth as an important source of satisfaction, with 38 percent of men checking this item.

Neither "your work itself" nor "your income" ranked very high with women as a source of satisfaction. The former, checked by 47 percent of women, was in fifth place as a source of satisfaction, ranking after marriage, children, social contacts, and avocational interests. Income, mentioned by only 15 percent of women, was at the bottom of the list.

There is no composite portrait to be made of the vocational careers of the gifted men and women for it is in this area that their many talents and great versatility are most evident. The men range in occupation from semiskilled labor to top-ranking university administrators, famed scientists, literary figures, high level officers and executives in business. The group is pretty well concentrated on the upper rungs of the vocational ladder with none at the bottom and only a few on the lower steps. But there is no evidence that the men with fewer vocational achievements are any less able intellectually than those who have reached high places. In some instances, the choice of vocation was determined by educational or occupational opportunities, in others by health, and in still others it was a matter of deliberate choice of a simple, less competitive way of life.

As for the gifted women, fewer than one-half are employed outside the home. Although, for most, a career is not of primary importance, a number of women have reached high levels of achievement. As a group, however, the accomplishments of the gifted women do not compare with those of the men. This is not surprising since it follows the cultural pattern to which most of the gifted women as well as women in general have succumbed. Not only may job success interfere with marriage success, but women who do seek a career outside the home have to break through many more barriers and overcome many more obstacles than do men on the road to success. Although the gifted women equaled or excelled the men in school achievement from the first grade through college, after school days were over the great majority ceased to compete with men in the world's work. This characteristic appears to be due to lack of motivation and opportunity rather than to lack of ability. Furthermore, an evaluation of achievement in terms of vocational accomplishment excludes the cultural contributions which the great majority of these women have made in many indirect and intangible ways and which perhaps are never properly evaluated.

In the following chapter we will describe some of the avocational interests of both men and women and discuss their participation in community life and their contributions to civic betterment.

CHAPTER VIII

AVOCATIONAL AND OTHER
INTERESTS

Among the characteristics that distinguished this group of subjects in childhood and youth was the breadth and versatility of their interests.[39] It is therefore interesting to investigate the extent to which the adult gifted subjects have continued to cultivate interests and activities not directly connected with their vocations. In order to obtain information of this kind the 1950 General Information Blank asked the subjects to list their avocational interests and hobbies and called also for information on memberships in clubs and organizations, and for a record of service activities including participation in community and civic affairs. In addition, the subjects were asked to report their publications, if any, as well as to describe any other creative work accomplished. The 1955 Information Blank, though it did not cover as many areas as the 1950 report, did bring up to date the information on publications and creative work of all kinds.

AVOCATIONAL INTERESTS

A wide variety of avocational interests and hobbies were mentioned by the 679 men and 510 women who supplied information on this item in the 1950 questionnaire. The average number of avocations or hobbies mentioned was 2.8 for men and 3.0 for women. About 5 percent of men and 3 percent of women reported no hobbies while 14 percent of men and about 17 percent of women reported as many as five or more avocational interests. These figures represent an increase over those of 1940 when 11 percent of men and 15 percent of women mentioned no hobbies or avocations. In the same period the proportion with four or more hobbies increased from 15 to 29 percent for men and from 21 to 34 percent for women. Age itself is no doubt a factor contributing to the greater interest in hobbies and avocations in the late thirties and early forties than had been shown ten or twelve years earlier when the subjects, at an average age of 29 were just getting launched on their careers. Some also were still unmarried in 1940 and more preoccupied with the opposite sex than with hobbies, while others, recently married, were busy with new homes and small children. The greater vocational and financial security of 1950 in the case of men and the freedom from the care of young children for many women made possible the devotion of more time to avocational pursuits. The percentage of men and women listing various numbers of hobbies in 1950 is shown in Table 43.

TABLE 43

Number of Hobbies (1950)

	Percentages	
	Men (N = 679)	Women (N = 510)
None	4.9	3.3
One	15.6	12.2
Two	27.4	23.1
Three	22.8	27.3
Four	15.3	17.3
Five or more	14.0	16.8

Although, as would be expected, there is considerable sex difference in the particular avocational interests and hobbies mentioned, it is interesting to find that the three leading avocations are the same for men and women though their rank order among the top three differs. The category of sports, which includes all active and participating sports and athletics as well as the spectator variety, was far ahead of all other interests for men with mention by 57 percent. In second place for men was music, followed closely by gardening. Among women, music ranked first as an avocation, reported by 45 percent; gardening was in second place, and sports ranked third. Workshop, in fourth place for men, is the expected counterpart of the domestic arts and handwork which ranks fourth among women. Only about 4 percent of women were interested in photography as a hobby as compared with 17 percent of men. On the other hand, the category "Art," which included the creative arts and art appreciation and also applied arts such as interior decoration, furniture design, house plans, et cetera, ranked fifth for women with 26 percent, and was mentioned as well by 11 percent of men. Table 44 lists those hobbies reported by as many as 10 percent of either men or women.

TABLE 44
Leading Hobbies (1950)

	Percentages	
	Men (N = 679)	Women (N = 510)
Sports	57.4	29.2
Music	33.1	44.9
Gardening	30.8	41.9
Home workshop activities	23.1	6.1
Photography	16.9	4.3
Art (creative and applied arts and art appreciation)	11.0	26.3
Creative writing	9.9	15.0
Social dancing (includes folk dancing, et cetera)	5.0	12.5
Domestic arts and handwork	2.3	28.5

Reading and Study

The information on hobbies tells only part of the story concerning the uses of leisure and means of relaxation. Of particular interest in a study of a group selected for intellectual superiority would be a detailed inquiry into the amount and kind of reading. Unfortunately our questions in this area were too general to yield information that can be quantified. No inquiry was made regarding amount of reading, and the phrasing of the question on reading preferences ("What kind of reading do you prefer?") makes it difficult to determine the favorite type of reading, since a large proportion of subjects listed more than one preference. However, the replies indicated that fiction, mentioned by more than two-thirds of the group, is by far the most popular kind of reading (and detective fiction ranks high as a favorite in this category). Biography, history, and travel are also favored but by a smaller proportion (35 to 40 percent). Nonfiction dealing with such topics as religion, philosophy, sociology, political affairs, or history, is the preferred reading of about 15 percent of the group and a preference for technical and scientific reading is mentioned by another 15 percent. Although the amount of reading was not called for, practically all subjects indicated a reading interest of some kind.

A number of subjects have continued to study informally. One-fifth of the men and one-fourth of the women reported study in a variety of fields either through independent reading or in groups such as the Great Books classes. In addition to literature, study in such fields as science, philosophy, and foreign languages was frequently mentioned.

SPECIAL ABILITIES AS RELATED TO VOCATIONS AND AVOCATIONS

Many of the subjects have displayed special abilities of various kinds. Talent along such lines as music, drama, art, mechanics, and writing were frequently reported by their parents and teachers. Often the special ability, if sufficiently marked, has determined the choice of vocation, but for a large number of subjects such talents have found their expression in avocational activities and hobbies.

Outstanding among the special talents noted is that of writing. In addition to the work of the career writers and journalists, a vast amount of material relating to their work has been published by men and women in scientific, technical, and other professional fields, but such professional output, even though the primary vocation is not writing, does not come under the heading of avocational writing. There are, however, between 125 and 150 subjects who write as a hobby or leisure-time activity. Of these, some 50 men and 40 women have had their work published. Although such writings are most often articles covering a variety of topics, they also include a number of short stories, poems, and plays as well as 15 nonfiction books and 8 or 10 novels, all the work of avocational writers. Especially noteworthy among these are the public relations executive, previously mentioned, whose first novel was a Book-of-the-Month selection, a banker who has published more than 25 short stories, the secretary who wrote a successful "western" novel, the army sergeant who writes science-fiction stories, and the housewife who has sold three short stories and is now revising a promising novel.

Much of the avocational writing has not been published, often done rather for enjoyment and self-expression and, of course, a great deal is not of the quality to merit publication. We can, however, look for more published material as the group grows older and as leisure time increases.

A number of other subjects have pursued their specialized gifts only as avocations. Among these are several gifted musicians. Musical talent, more often than any other gift, has been turned to use as a means of livelihood while preparing for another vocation. One man, for example, who has become a research scientist, gave evidence at an early age of marked musical ability and for some years he was undecided between music and science as a life work. He began concert work at the age of 14 and later became a conductor of a civic orchestra. His musical talent enabled him to finance several years of graduate study and now provides a satisfying avocational outlet.

A good many of those with dramatic ability are participating in such activities as the "Little Theater" or "Children's Theater" or other community dramatics. Some are actors while others have engaged in stage production or direction in amateur productions.

Eight or ten members of the group have displayed above-average ability in art. Among these are four schoolteachers whose paintings have appeared in various exhibits and have been highly praised by critics. There are also several amateur photographers whose fine camera work has won awards in competitive shows.

Many other examples of unusual talent along specialized lines could be given, but the following two illustrations, while not typical, will serve to give some idea of the versatility to be found in the group.

A woman who by the age of 40 has had several careers. Her first success was in acting. As a child she appeared in motion pictures, then was for several years a professional actress in major theatrical productions both in London and in New York. She was also a professional dancer and an amateur championship ice skater. After giving up her professional acting career she continued her work in dramatics as an avocation, working chiefly with children's theatrical groups. She also possessed unusual talent in drawing and was commissioned to do the illustrations for two textbooks, one in anatomy and one in physiology. She later entered the field of business where she has an executive position. She has written several plays that have been produced by amateur groups and has also written two novels. Though neither of the novels has been published, both have received favorable comment and one is being considered for publication.

A man who is a lawyer maintains a limited and highly specialized law practice in order to have time for his special interests. Among these are research in stereoscopic optics which grew out of his work in photography. His research in this field of science is believed so important and valuable that the government is very much interested in the results of his project. He is also gifted in languages and is proficient in both German and French. He does considerable translating, mostly of scientific, technical, and legal articles for French and particularly German publications, and has also written several original articles and poems published in German periodicals. More recently, he has specialized in the Arabic language. He not only reads extensively but has also taught courses in Arabic. Military affairs are still another avocation, and he is especially active in the National Guard.

In addition to being a officer in the National Guard, he has prepared instructional material of various kinds, including the development of audiovisual training aids. He also contributes articles from time to time to military journals.

MEMBERSHIPS

An indication of the social inclinations and group interests of the gifted adult is to be found in the number and kinds of memberships reported in 1950. Since a number of the affiliations were with business or professional and service groups, it is not surprising that a larger proportion of men than women belonged to one or more clubs or organizations. Less than 15 percent of men listed no club or organization memberships but 31 percent of women had no organized group affiliations. The figures on the number of memberships in clubs or organizations are as follows for the 680 men and 521 women who replied to this item:

Memberships	Men %	Women %
None	14.7	31.5
One	20.0	27.3
Two	19.8	19.4
Three	14.6	11.7
Four	14.3	4.4
Five or more	16.6	5.7

Four or more memberships were reported by 31 percent of men and 10 percent of women. The organizations were classified according to type of activity or interest. The "kinds" of memberships are loose groupings with considerable overlap in function. Social overtones are found in all the groups and, certainly, service activities are not limited to the so-called service organizations. Many of the essentially social groups adopt a cause and doubtless many men and women join a service club for social or business reasons. The classification of church-related organizations does not include church membership, which will be discussed later in this chapter. Rather, it refers to groups made up of adherents of particular faiths, e.g., the Women's Society for Christian Service, B'nai B'rith, Knights of Columbus, and other groups that combine religious activity and service functions as well as promoting social relationships. The foregoing are examples of the overlapping to be found in most of our membership groups with the exception of the military organizations. The political and government category is also perhaps a more discrete grouping than some of the other categories. It includes organizations such as the League of Women Voters, Good Government League, American Civil Liberties Union, World Affairs Council, the NAACP, the Zionist Organization of America, United World Federalists, Council for Civic Unity, Americans for Democratic Action, and so on. Table 45 gives the proportion of men and women affiliated with each type of organization.

TABLE 45

MEMBERSHIPS IN CLUBS OR ORGANIZATIONS (1950)

Type of Organization	Percentages*	
	Men (N = 680)	Women (N = 521)
Business, professional and kindred.....	67.2	23.6
Social and fraternal..................	39.2	39.9
Recreational and hobby	15.4	10.8
Service (Kiwanis, Rotary, Soroptomist, et cetera)	11.6	4.8
Study and cultural	8.4	9.6
Politics and government..............	5.9	8.8
Church-related	4.9	9.8
Military (Reserve Officer, National Guard, et cetera)	4.7	0.4

* The percentages will not add to 100, since (1) the same individual may appear in more than one category, and (2) 15% of men and 31% of women were not members of any organized group.

The business, professional, and other job-related affiliations were by far the most frequent type of membership for men. Two-thirds of the men belonged to at least one organization related to their occupation. This category included such professional and business groups as the American Medical Society, the Bar Association, the Pasteur Society, the Underwriters' Club, National Association of Cost Accountants, Western Society of Naturalists, American Rocket Society, the Industrial Relations Research Association, and the Chamber of Commerce and similar business groups, to mention only a few of the many such organizations. The category also included membership in trade unions or guilds to which at least 65 men and 15 or 20 women belonged. Among the most frequently mentioned labor organizations were the guilds and unions connected with entertainment media such as radio, television, motion pictures, and the theater. These included the Musicians Union, Radio Writers Guild, Dramatists Guild, Radio and Television Directors Guild, Cameraman's Union, and so on. Other organizations represented by five or more members were the newspaperman's guild and various office worker unions. Membership in several other guilds and trade unions was reported but there were no more than two or three in any one of these.

Several subjects have been active in the labor movement and have held official positions in their unions or in the AF of L or CIO. Prominent among these is a man with a Ph.D. in economics who holds the position of Economic Counsel, directing research and negotiations for a trade union. He is a member of the American Economic Association and the American Statistical Association as well as of the Labor Council of the city in which he works. Two men have been members of a local CIO Council, another has been editor of the union paper, and still others have held office or done committee work in their organizations.

Groups organized primarily for social purposes are the second most frequent type of membership reported by gifted men, while social groups rank first with the women. Approximately two-fifths of both men and women belong to one or more clubs, lodges, or other social group. Recreational and hobby clubs ranked third in popularity with both men and women, though a larger proportion of men than women mentioned such groups. The following are examples of some of the groups in this category: Garden Club, Little Theater Group, Sports Car Club, National Rifle Association, Camera Club, Folk Dancing Club, Mountain Climbing Club, and Choral Society, to mention a few.

Only the "service clubs" among the remaining categories were claimed by as many as 10 percent of the men. No other group had a 10 percent representation among women, although both the church-related groups and the study and cultural organizations were close with 9.8 and 9.6 percent respectively.

Service Activities

Although the data on memberships are interesting, they present only one aspect of the gifted adult as a participating member of society. To complete the picture, we turn to the subject's report on service activities. This item in the 1950 General Information Blank was worded as follows: *Record of your service activities (such as scout work, welfare, religious or church work, participation in community and civic affairs, P.T.A., etc.).*

The memberships in Table 45 included only private organizations with formal membership rolls, as opposed to such community- or nationwide groups as P.T.A., Red Cross, American Cancer Society, Boy Scouts (or Girl Scouts), the Y.M.C.A., or any of the other civic or community welfare groups. Nearly one-half of the men and two-thirds of the women were engaged in one or more of the activities sponsored by such groups; in fact, 10 percent of men and nearly 22 percent of women mentioned 3 or more community services. In view of their age and family status, it is not surprising that the most frequent interest of both men (22%) and of women (48%) was in organizations concerned with youth. Prominent among these were the P.T.A., the Scouts, and the "Y." About 15 percent of men and 10 percent of women have served with community health organizations (e.g., Mental Health Society), on school boards or as school trustee, in civilian defense programs, city planning boards, and with similar organizations. Service activities in connection with a church or of a religious nature were reported by 13 percent of men and 19 percent of women. Volunteer welfare work with the Red Cross, Crippled Children's Society, Heart Association, Community Chest, Alcoholics Anonymous, Grey Ladies, and various similar projects was listed by 11 percent of men and 22 percent of women.

While the reports on service activities indicate in some degree the extent of participation of our gifted group in community life, they do not tell the full story. Some subjects are more modest than others or attach less importance to their work with the result that their reports on this phase of their lives are too general for classification. However, even the minimal figures available are impressive evidence of the social awareness of the gifted and their willingness to contribute to community life. Through membership on boards or in groups that make community policy and through other volunteer activities, especially in the case of women, a great deal of worth-while work of both "staff" and "line" variety is accomplished.

A number of the subjects have received public recognition and honor for their contributions to community welfare and public betterment, and some illustrations follow. Among the men there have been: Man of the Year (large western city); Recognition Medal for Distinguished Community Leadership (large western city); Outstanding Citizen (county award); four men with Distinguished Civilian Service Awards; a Special Award for Distinguished Service to Boyhood; three listings in America's Young Men—to mention some of the more outstanding. Too numerous to cite are those who have held office in the various community service or civic betterment organizations. Included here are such positions as membership on the Board of Directors of Jewish Big Brothers; chairman of Budget Committee of Community Chest (metropolitan area); member, Board of Directors of a city chapter of the American Cancer Society; chairman of a County Planning Commission; director of the Legal Aid Society; and president, Youth Welfare League. The list of women who have won recognition or held important posts in community organizations is long. Some examples follow: Special Service Award (to four women); Woman of the Year (in education); president of area Girl Scout Council; president of Women's Board, Museum of Art; appointment to the Board of Education of a large city; chairman of State Committee of League of Women Voters; and appointment to a Grand Jury. The foregoing are illustrations from a long list of offices held and citations received by members of the gifted group, both men and women.

The Religious Outlook

Although we inquired into the attitudes and interests of the subjects with the results presented above, the questions were not specific so far as particular hobbies, interests, or activities were concerned. An exception to this was in the matter of religion. The Supplementary Biographical Data blank of 1950–52 asked for the amount of religious training in youth, the extent of religious inclination as an adult, and the religious affiliation. Close to three-fifths of both men and women reported that they received "very strict" or "considerable" religious training; somewhat more than one-third reported "little," and about 6 percent said they had no religious training.

As adults, 38 percent of men and 53 percent of women expressed moderate to strong religious inclinations. The 599 men and 491 women for whom data are available responded as follows:

Religious inclination	Men %	Women %
Strong	10	18
Moderate	28	35
Little	34	24
None at all	28	23

With respect to religious affiliation, 59 percent of men and 56 percent of women say they belong to a particular church, congregation, or other religion-oriented group. These figures approximate the 57 percent of the total population (sexes are not reported separately) reported as church members in 1950.[45] Among the gifted an additional 8 percent of each sex, while not formally affiliated with any church or religious group, say that they attend services or are inclined toward a particular faith. A small group of about 4 percent of men and 2 percent of women state that they have a personal faith or are interested in the philosophy of religion but that creeds and denominations or other forms of organized religion have no appeal for them. Almost 24 percent of the men and 32 percent of the women reported that they are not affiliated with any religious group and made no comment regarding their attitude toward religion. A small minority (5% of men and 1% of women) described themselves as skeptics, agnostics, or, in a few cases, as atheists.

The sex difference in church membership, though small (59% of men versus 56% of women), is of interest because it is not in the expected direction, since most investigations show that more women than men are church members. Havemann and West[17] report that among college graduates about 10 percent more women than men are church goers.

SUMMARY

More than four-fifths of the subjects reported an interest in two or more avocational pursuits and more than one-half reported three or more. The absence of any norms precludes a comparison with the general population in this matter, but the data indicate a considerable breadth and diversity of interests.

Many of the special abilities that had been evidenced by the subjects in their youth found expression in hobbies and avocations at mid-life. Talent in creative writing, art, dramatics, and music was especially noteworthy.

Complete data on the amount and kinds of reading and on reading interests are lacking. The information available, however, indicates wide reading interests, covering many fields of fiction and nonfiction. Study through independent reading or in informal groups and classes was also mentioned by a number of subjects.

Four-fifths of the men and two-thirds of the women reported membership in one or more clubs or organizations, chiefly professional or business and social.

The group has displayed an interest in and responsibility for the community and civic welfare through participation in a wide variety of activities such as organizations concerned with youth, health programs, civic betterment projects, and similar plans. Outstanding contributions along these lines have won recognition and special awards for a number of men and women.

An inquiry regarding religious interests elicited the information that somewhat less than two-fifths of the men and more than one-half of the women feel moderately or strongly inclined toward religion ; however, in the matter of religious affiliation, the men report a slightly higher proportion of church memberships.

CHAPTER IX

SOME POLITICAL AND SOCIAL ATTITUDES

The General Information Blank of 1950 called for information on political affiliation, voting habits, and a self-rating on radicalism-conservatism. Since similar information had been secured in the 1940 follow-up, it is possible both to picture the 1950 political and social attitudes and to compare the 1950 attitudes with those expressed in 1940,[36] to learn what changes may have occurred in the ten-year interval.

SELF-RATINGS ON RADICALISM-CONSERVATISM

The self-ratings on radicalism-conservatism (r-c ratings) employed a cross-on-line technique in which the rating bar represented a continuum ranging from "extremely radical" at one end to "very conservative" at the other. The responses were evaluated on a nine-point scale. The extreme left of the horizontal bar, defined as "extremely radical," was coded *1*; "tend to be radical," *3*; "average," *5*; "tend to be conservative," *7*; and "very conservative," at the extreme right, *9*. The intervening even numbers, *2, 4, 6,* and *8,* represented the midvalues between the adjacent categories. The directions were simple: "Rate yourself on the following scale as regards your political and social viewpoint." Only rarely did a subject complain of ambiguity in the question, although an occasional respondent checked himself at one level on political and at another on social viewpoint. For all but a very few, the ratings presumably represented a composite of general attitude on political and social issues based, of course, on individual concepts of the term "political and social viewpoint" and of what constitutes the "average" in this regard.

A total of 1,241 subjects (698 men and 543 women) rated themselves on this variable in 1950. The distributions of the r-c ratings with means and standard deviations are given in Table 46.

TABLE 46

Self-Ratings on Radicalism-Conservatism (1950)

	Rating	Men N	Men %	Women N	Women %
1.	Extremely radical	1	0.1	...	...
2.		9	1.3	5	0.9
3.	Tend to be radical...........	62	8.9	56	10.3
4.		93	13.3	76	14.0
5.	Average	199	28.5	189	34.8
6.		82	11.8	56	10.3
7.	Tend to be conservative......	206	29.5	140	25.8
8.		29	4.2	12	2.2
9.	Extremely conservative	17	2.4	9	1.7
	Total	698		543	
	Mean rating	5.57		5.38	
	Standard deviation	1.54		1.45	

According to their own opinions, more than half the men and nearly three-fifths of the women consider themselves "middle-of-the-roaders" though the inclination is slightly to the right of center. The men tend to be more conservative than the women, with a mean rating of 5.57 as compared with a mean of 5.38 for women. The sex difference in mean ratings is fairly reliable $(P = .03)$. The differences between men and women are found chiefly in the proportions at the average or near-average categories (ratings *4, 5, 6*) and at the conservative end of the scale (ratings *7, 8, 9*). Somewhat more women than men rate themselves average (59% versus 54%) and more men (36%) rate themselves conservative than do women (30%). The proportions to the left of center are not very different for the sexes: 10% of men and 11% of women rate themselves *1, 2,* or *3*.

There were 602 men and 494 women who rated themselves on radicalism-conservatism in both 1940 and 1950. Table 47 compares the two sets of ratings made by the same subjects ten years apart.

TABLE 47

Comparison of Self-Ratings on Radicalism-Conservatism
of 1940 and 1950

	Men 1940 N	Men 1940 %	Men 1950 N	Men 1950 %	Women 1940 N	Women 1940 %	Women 1950 N	Women 1950 %
Ratings								
On radical side (*1, 2, 3*)......	136	22.6	62	10.3	91	18.4	57	11.5
Average or near average (*4, 5, 6*).................	289	48.0	325	54.0	282	57.1	283	57.3
On conservative side (*7, 8, 9*)	177	29.4	215	35.7	121	24.5	154	31.2

Ratings at the radical end of the scale for these twice-rated people decreased from almost 23 percent to 10 percent for men and from 18 percent to 11 percent for women between 1940 and 1950. The differences in percentages at the two dates are statistically significant for both sexes ($P = <.001$). At the other end of the scale the conservative ratings increased from 29 to 36 percent for men and from 24 to 31 percent for women. Ratings at both extremes of the scale, however, were less frequent in 1950. There were eight men and four women who ranked themselves at *1*—the point farthest to the left—in 1940, but no one checked this point in 1950. At the far right, 18 men checked *9*, extremely conservative, in 1940 as compared with only 13 in 1950.* However, of seven women who rated themselves *9* in 1940 and eight who did so in 1950, only one was the same individual. Six of the eight had been at the "'tend-to-be-conservative" point in 1940 and one had rated herself at the far left in 1940!

Even though the general trend between 1940 and 1950 was toward greater conservatism, there were shifts in both directions, and there were also a considerable number of persons whose position on the r-c scale remained unchanged. Identical ratings were given at both dates by 37 percent of men and 40 percent of women. Shifting to a more radical point on the scale were 19 percent of men and 21 percent of women, and shifting toward greater conservatism were 44 percent of men and 39 percent of women. The correlation between the self-ratings of 1940 and 1950 was .62 for both men and women.

VARIABLES ASSOCIATED WITH 1950 R-C RATINGS

It is interesting to examine the relationship between the political and social attitudes and certain other characteristics. A discussion of some of the variables associated with the r-c ratings follows.

Age.—That age is to some extent a factor in the trend toward conservatism is shown in Table 48, which gives the mean r-c ratings according to the year of birth. The youngest group of men (those born in 1915 or later) averaged reliably less conservative than those who were older, with a mean r-c rating of 5.18 for the youngest as compared with a mean of 5.62 for the total born before 1915. The three older

---

* It should be remembered that the numbers referred to in this paragraph are for subjects in the twice-rated sample. Table 51 gives the total number of *1* and *9* ratings in 1950. In 1940, of the total 667 men with r-c ratings, 11 gave themselves a rating of *1* and 21 considered themselves to be *9*. Of the total 543 women with 1940 ratings, four rated themselves *1* and nine rated themselves *9*.

groups of men do not differ to any extent in mean r-c rating. The same tendency toward greater liberalism among those born in 1915 or later is found for women but the difference is less marked than for men and is not statistically significant. The mean r-c rating for the total group of women born before 1915 is 5.41 compared with 5.16 for the youngest group. It is interesting to observe, however, that the 15 oldest women are far more conservative than any other group.

TABLE 48

1950 SELF-RATINGS ON RADICALISM-CONSERVATISM BY YEAR OF BIRTH

	Men			Women		
Year of Birth	N	Mean r-c rating	S.D.	N	Mean r-c rating	S.D.
Before 1905......	58	5.62	1.48	15	6.27	1.12
1905–1909	228	5.61	1.59	177	5.39	1.45
1910–1914	335	5.64	1.53	277	5.38	1.46
1915 or later.....	77	5.18	1.38	74	5.16	1.39

Education.—Among both men and women, the most conservative are those who entered college but did not graduate. The college graduates are nearer the center (the women more so than the men) and those who did not attend college at all fall between the other two educational groups. Within the college-graduate group the men show differences in r-c rating according to academic degree. The range in ratings is from a mean of 5.03 for the 68 men with a Ph.D. to 5.75 for the 81 LL.B.'s, and for the total group of 280 men with a graduate degree the mean is 5.46. The mean r-c rating for men with a bachelor's degree only is 5.72. For the 123 women with a graduate degree the mean r-c rating is 4.95 in comparison with 5.54 for the women with only a bachelor's. Because the number of women involved was relatively small, the mean r-c ratings according to the various graduate degrees were not computed. Table 49 gives the r-c ratings according to educational level for both men and women.

TABLE 49

1950 SELF-RATINGS ON RADICALISM-CONSERVATISM BY AMOUNT OF EDUCATION

	Men			Women		
	N	Mean r-c rating	S.D.	N	Mean r-c rating	S.D.
1. College graduates.	504	5.51	1.54	383	5.30	1.41
2. College 1 to 4 years	107	5.83	1.44	78	5.60	1.44
3. No College	87	5.62	1.60	82	5.54	1.49

Occupation.—As might be expected, political and social viewpoint varies according to occupation. Men in the professions (Group I of the Minnesota Occupational Scale) have a mean r-c rating of 5.40 as compared with 5.72 for Group II, the semiprofessional and managerial occupations. Men in Group III (clerical, retail business and skilled trades), with a mean of 5.73, rated themselves almost exactly the same as those in the higher echelons of business (Group II). The most conservative group were the 11 men in Group IV (agriculture and related occupations) whose mean r-c rating is 6.45. There are some marked differences within the occupational groups which are shown in Table 50 where the rank order of 21 occupations from most liberal to most conservative is given.

TABLE 50

1950 SELF-RATINGS ON RADICALISM-CONSERVATISM
BY OCCUPATION (MEN)*

(from most liberal to most conservative)

	Occupation	N	Mean r-c rating	S.D.
1.	Personnel directors or welfare workers.......	11	4.27	...
2.	Authors or journalists	17	4.29	1.27
3.	Clergymen	7	4.43	...
4.	Economists or political scientists............	7	4.43	...
5.	Entertainment (directors, producers, writers).	14	4.71	1.52
6.	College or university faculty	40	4.75	1.39
7.	Teachers below college level................	35	5.08	1.21
8.	Architects	8	5.38	...
9.	Accountants or statisticians.................	38	5.47	1.48
10.	Chemists or physicists.....................	32	5.50	1.46
11.	In advertising, publicity, or public relations...	22	5.64	1.32
12.	In clerical, sales and retail business...........	34	5.65	1.80
13.	In skilled trades...........................	25	5.68	1.30
14.	Physicians	41	5.73	1.15
15.	Executives in business and industry..........	59	5.85	1.24
16.	Lawyers	71	5.85	1.41
17.	Engineers	48	5.88	1.76
18.	Army or Navy officers......................	16	6.19	1.47
19.	Banking, finance, or insurance executives.....	36	6.39	1.28
20.	Farmers and ranchers......................	11	6.45	...
21.	Sales managers, sales engineers, or technical salesmen	19	6.58	1.23

* Occupations in which fewer than seven men are engaged are not reported separately.

Despite the fact that for three of the occupations in Table 50 the number of men rated is less than 10, the rank order of the occupations seems, in general, fairly plausible. Most persons would agree that men in occupations ranking 1 to 6 are usually more liberal in their political attitudes than men in occupations 18 to 21, or than those in positions 11 to 17. Some may be surprised, however, that men in skilled trades should rank 13 in the list—only two ranks less conservative than executives in business or industry, and three ranks less conservative than lawyers; or that farmers should rank as one of the four most conservative groups.

Too few women were employed to permit an analysis by particular occupation, but a comparison of broader vocational groupings reveals some interesting differences in r-c ratings. The most liberal group are the women in the miscellaneous professional occupations (see Table 31), for whom the mean rating is 4.63. Women on college faculties and in the higher professions, and the school teachers are both slightly to the right with a mean rating of 5.17. The most conservative are the business women and office workers whose average r-c rating is 5.64. The housewives, who make up by far the largest group with a total of 339 self-ratings, have a mean r-c rating of 5.50. Table 51 gives the r-c ratings for women by occupation.

TABLE 51

1950 Self-Ratings on Radicalism-Conservatism
by Occupation (Women)

(*from most liberal to most conservative*)

Occupation	N	Mean r-c rating	S.D.
Miscellaneous professional occupations	54	4.63	1.40
College faculty and higher professions	29	5.17	1.70
Schoolteachers (including administrators and counselors)	46	5.17	1.46
Housewives*	339	5.50	1.33
Office, business, and clerical occupations..........	66	5.64	1.51

* Housewives who have full-time jobs outside the home are included in the appropriate occupational category and do not appear here.

Income.—Surprisingly, there was little relationship between the 1950 r-c ratings and earned income. For men the largest difference was between the 322 with earned incomes of $7,000 and above in 1949 and the 330 who earned less than $7,000. The former were the more conservative with a mean rating of 5.72, while for the latter the mean was

5.45. The difference between the r-c means is fairly reliable ($P = .02$). For women, the trend was reversed. The 21 women who earned $6,000 or more were the most *liberal* of all the income groups by a slight but unreliable margin. When mean r-c ratings were checked against the total annual income of husband and wife, whether earned or unearned, no significant relationship was found. However, both the gifted men and gifted women whose total family income was in the upper half (above $7,000) rated themselves slightly more conservative than those whose incomes fell below $7,000.

Intelligence.—One would like to know how the r-c variable is related to intelligence in the general population. Our data do not answer this question since our gifted subjects as adults score far above the generality of adults on the Concept Mastery test and were, by definition, in the top one percent of the generality in childhood IQ. However, despite the restricted intellectual range in our gifted group, Table 52 shows that radicalism (or liberalism) is positively associated with higher scores on the Concept Mastery test in 1950. Men who rated themselves as *1, 2,* or *3* have a mean CMT score of 155 compared with a mean of 140* for men who rate themselves as *4, 5,* or *6.* The difference is highly reliable ($P = .001$). The corresponding drop in the CMT score for women, from a mean of 153.7 for those on the radical side to 132.7 for those in the center, is even more significant ($P = .0001$). Those who rated themselves as conservatives have the lowest CMT scores but the drop in mean score for ratings *7, 8,* or *9* as compared with ratings of *4, 5,* or *6* is not very reliable for either sex ($P = .09$ for men and .05 for women).

Additional evidence of the tendency for those who score highest on intelligence tests to be less conservative is found in the data for the Stanford-Binet IQ's. Although the number of cases with IQ's of 170 or over is too small to warrant conclusions, it is of interest to find that the 42 men with childhood IQ's of 170 and above who rated themselves on radicalism-conservatism in 1950 have a mean rating of 5.07 (S.D. 1.80) as against a mean of 5.57 for the total group of men. For the 31 women whose Binet IQ's were 170 and above in childhood, the 1950 mean r-c rating is 5.19 (S.D. 1.51) in comparison with a mean of 5.38 for all gifted women. Although the subjects with the highest childhood IQ's are more liberal on the average than other members of the group,

* These are "point" scores, not IQ's.

they, too, grew more conservative in the interval between 1940 and 1950. The r-c mean in 1940 for this selected group of highest IQ was 4.83 for men and 4.90 for women.[36]

TABLE 52

1950 SELF-RATINGS ON RADICALISM-CONSERVATISM
BY SCORES ON CONCEPT MASTERY TEST

| | Concept Mastery Scores | | | | | |
| | | Men | | | Women | |
Ratings	N	Mean	S.D.	N	Mean	S.D.
On radical side (1, 2, 3).........	49	155.1	21.7	50	153.7	15.8
Average or near average (4, 5, 6)	280	140.1	28.6	248	132.7	29.0
On conservative side (7, 8, 9).....	179	135.3	29.0	116	126.8	25.9

General adjustment.—A comparison of the ratings on mental health and general adjustment* shows that those subjects classified as having some, or serious difficulty in adjustment rated themselves reliably more liberal on the average than those considered satisfactory in adjustment. Table 53 gives the mean r-c rating according to general adjustment classification. The differences in mean r-c rating between those rated satisfactory in general adjustment and those with some, or serious difficulty (categories 2 and 3) are statistically significant ($P = < .001$ for both sexes).

TABLE 53

1950 SELF-RATINGS ON RADICALISM-CONSERVATISM BY
GENERAL ADJUSTMENT CLASSIFICATION

| | Men | | | Women | | |
General Adjustment	N	Mean r-c rating	S.D.	N	Mean r-c rating	S.D.
1. Satisfactory	511	5.72	1.49	375	5.54	1.34
2. Some difficulty	136	5.22	1.50	133	5.13	1.58
3. Serious difficulty	51	5.20	1.76	35	4.60	1.63

The tendency for the less well-adjusted to rate themselves as more liberal in political and social viewpoint was also found in 1940. At that time, however, these subjects were farther to the left with mean r-c ratings of 4.61 and 4.64 for men and women, respectively.

It was pointed out in Chapter IV that the subjects who experience difficulty in emotional and social adjustment make reliably higher scores on the Concept Mastery test. As shown above, the subjects to the left

* See Chapter IV for a description of these ratings.

of center on the r-c scale also have significantly higher scores on the Concept Mastery test. Thus, we have a situation in which those more liberally inclined, socially and politically, make significantly higher test scores (Concept Mastery) and are also more often classified as having some, or serious difficulty in general adjustment. We might theorize that the higher Concept Mastery scores of the less well-adjusted may be due to their relative lack of social ability, their tendency toward more solitary interests, and consequent greater preoccupation with intellectual pursuits. The corresponding tendency to liberal or radical attitudes found for the less well-adjusted may perhaps be explained by the lack of conformity that contributes to the adjustment difficulties.

Although the differences disclosed in Tables 52 and 53 are statistically reliable, care should be taken against overgeneralization. Actually the differences in mean r-c ratings according to general adjustment are much too small to justify the conclusion that conservatives are usually well-adjusted and that radicals usually are not. Nor in view of the small number of cases at the radical end of the scale can it be concluded that those who rate themselves left of center are more intelligent, as measured by test scores, than the conservative members of the group.

In interpreting the data on political and social attitudes, particularly with reference to radicalism-conservatism, it should be borne in mind that these opinions were expressed chiefly in 1950 and 1951. If a similar opinion survey of the group were made now, the results might be different, not alone because of the influence of age, but because of the changes in the political and economic scene and in the concept of what constitutes radicalism or conservatism.

POLITICAL AFFILIATIONS

Information on political preferences was supplied by approximately 1,250 subjects in response to the question: *On national issues which of the political parties most nearly represents your leanings? (Check) Democrat —— Republican —— Socialist —— Communist —— Other (Specify) ——.* Somewhat more than one-half (55%) of the men and one-half of the women said they were Republicans. One-third of the men and two-fifths of the women said they were Democrats and 2.6 percent of each sex identified themselves as Socialists. No one claimed membership in the Communist party. Not aligned regularly with any of the major parties were 7.8 percent of men and 6 percent of women. Most of the latter described themselves as "Independent" and not fitting

a party pattern, though a few said "None." Slightly more than one percent of each sex mentioned a minor party, usually the Independent Pro-
gressive Party or other liberal group.

In the 1940–50 decade the political affiliations showed the same
trend toward greater conservatism as had been shown by the r-c ratings.
About 10 percent more men and 8 percent more women called themselves Republican in 1950 than in 1940. The number of Democrats
among men decreased from 40 percent in 1940 to the 33 percent of 1950
while the proportion of women under the Democratic banner changed
hardly at all. The Socialists in each sex decreased from approximately
4 percent in 1940 to 2.6 percent in 1950. At the earlier date a larger
proportion (10 percent of men and 12 percent of women) were in the
"None" or "Independent" category.

A comparison of party affiliation with r-c self-rating shows the
Democrats, both men and women, to be reliably more liberal in political and social viewpoint than were the Republicans. The Democrats,
though slightly left of center, were much closer to the middle than were
the Republicans. Among men, the Democrats with a mean r-c rating
of about 4.51 were only a half-step from the middle point on the scale,
while the Republicans who averaged 6.48 were nearly one and one-half
steps to the right of center. The women showed the same trend though
the contrast was not quite so marked: Democrats averaged 4.60 on
the r-c scale and Republicans 6.25. Those with no affiliation were only
slightly to the left of center with a mean 4.91 for men and 4.88 for
women. Table 54 gives the percentage distribution of political party
preferences and includes also the mean self-rating on radicalism-conservatism according to political preference.

TABLE 54

POLITICAL PREFERENCES AND RELATIONSHIP TO
RADICALISM-CONSERVATISM (1950)

	Men			Women		
Political Party	N	%	Mean r-c rating	N	%	Mean r-c rating
Republican	387	55.1	6.48*	268	49.5	6.25*
Democrat	233	33.1	4.51	219	40.5	4.60
Socialist	18	2.6	3.47	14	2.6	3.07
Other	10	1.4	2.88	7	1.3	3.57
No affiliation	55	7.8	4.91	33	6.1	4.88

* The critical ratio of the difference in mean r-c rating between Republicans and Democrats
is 21.9 for men and 16.5 for women.

VOTING HABITS

The General Information Blank of 1950, like that of 1940, asked for the voting habits of the subjects with respect to national, state, and local elections. Replies to this item were received from 690 men and 531 women. The responses classified in Table 55 show that those voting "always" or "usually" include approximately 94 percent of men and 98 percent of women in national elections, 90 percent of men and 95 percent of women in state elections, 85 percent of men and 91 percent of women in local elections. These percentages are enormously higher than those reported for the general population, about 60 percent of whom vote in national elections. The percentages of gifted subjects voting "usually" or "always" was about the same in 1950 as it had been in 1940. One can only conclude that these gifted people as a rule take their civic obligations much more seriously than is true of the general population.

TABLE 55

VOTING HABITS (1950)

	Percentage Voting					
	National Elections		State Elections		Local Elections	
	Men	Women	Men	Women	Men	Women
Always	82.0	89.4	70.2	75.1	49.2	58.7
Usually	12.0	8.7	20.4	20.0	36.2	32.6
Occasionally	1.9	0.2	3.2	1.9	6.5	4.5
Rarely or never	4.1	1.7	6.2	3.0	8.1	4.2

POLITICAL ACTIVITIES

Several men in our group have sought public office, some successfully, some not. Among the elective offices held are two superior court judgeships (one of these men has since been appointed to the appellate court), one municipal court judgeship, two members of the state legislature and one high state official whose name has been mentioned as a possible United States senator. Another was defeated in a campaign for superior court judge, two have been unsuccessful candidates for the State Assembly and one man was defeated in his candidacy for the United States House of Representatives. Membership on the state central committee of either the Democratic or Republican party has been reported by several men and still others have served on county central committees of the major political parties. Altogether, a number of subjects, both men and women, have engaged in political campaigns as committee members, speakers, precinct workers, and other activities.

Others have held appointive offices, both federal and state. Among these are two lawyers who are judicial referees in state agencies. Still others have held important administrative positions in agencies of the federal government. One of the most important appointive positions was held by a young man who, while still under 33 years of age, was made an administrative assistant to the President of the United States and served as a member of the White House executive staff for nearly seven years. He won a Rockefeller Public Service Award for outstanding service in government.

Our records are not complete on the extent of participation of the group in political affairs, since no specific information on that type of activity has been sought since the 1950–52 follow-up. Appointment or election to public office was reported in the 1955 mailed questionnaire but minor activities were not called for and, therefore, were seldom mentioned. Because of the incomplete nature of the present data and because of the relative youth of the subjects, a record of greater participation in public life and more political activity can be looked for in future reports.

SUMMARY

The men and women of the gifted group at mid-life consider themselves close to the center on a radicalism-conservatism continuum. Both sexes are slightly to the right of the mid-point on the nine-point scale on which they rated themselves. In political affiliation, somewhat more than half of the men and about half of the women are Republicans. The group as a whole has a remarkable voting record, as shown by the better than 90 percent who report that they "usually" or "always" vote in state and national elections.

A comparison of the self-ratings on radicalism-conservatism made in 1950 with those made in 1940 shows a shift in the direction of less radicalism and greater conservatism. Although the increase in proportion at the more conservative end of the scale from 1940 to 1950 was statistically reliable, it should not be overlooked that about one-fifth of the group had moved closer to the liberal end of the political spectrum in 1950. The changes in political affiliation from 1940 to 1950 are in the same direction as the r-c ratings ; that is, in 1950 more labeled themselves as Republicans and fewer as Democrats than in 1940.

Several members of the group have held important elective offices and several others have been appointed to responsible positions in both state and federal government. Since the group is still relatively young, greater participation in political life and government can be expected.

An examination of the political and social attitudes in the light of certain other variables reveals a relationship between r-c ratings and age, extent of education, and certain occupations. There is also a tendency for ratings of *1, 2, 3* on the r-c scale (tend to be radical) to be associated with higher Concept Mastery test scores as well as with poorer general social and emotional adjustment.

CHAPTER X

MARRIAGE, DIVORCE, AND OFFSPRING

Marriage and marital adjustment are important factors in a portrayal of anyone's life success and happiness. How do our gifted subjects compare in these respects with the generality of men and women of corresponding age?

Consider first the incidence of marriage to 1955 when the average age of the subjects was approximately 44 years. By that time 93 percent of the men and almost 90 percent of the women had married. These percentages are based on the 780 men and the 610 women living in 1955 for whom information on marital status was available. The incidence of marriage for both sexes is about the same as that reported for the total population of corresponding age. In other words, being highly intelligent apparently is not an obstacle to marriage for either sex, at least that is true for this group. Table 56 gives the proportion at specified age levels who have ever been married regardless of current marital status. The average age at marriage (first marriage if more than one) was 25 years for men and 23 years for women. Twenty-three men (3%) married before reaching age 21 and 9 of these were 18 when married. The number of women married before age 21 was 71 (12%) and of these 19 were married at ages 16 to 18. On the other hand, six men and nine women did not marry until after age 40. Among the men, 45 has been the oldest age at first marriage, and for women it was 51 years.

TABLE 56

INCIDENCE OF MARRIAGE ACCORDING TO AGE

Age	N*		Percentage Who Have Married	
	Men	Women	Men	Women
Under 40	83	74	96.4	93.2
40–44	361	316	91.7	90.5
45–49	262	200	93.5	87.5
50 and over...........	74	20	93.2	80.0
Ages combined	780	610	93.0	89.5

* The N's here do not include 4 men and 2 women for whom information on 1955 marital status was not obtained; nor do they include the 11 men and 17 women lost since 1928 or earlier.

132

A comparison of the incidence of marriage among the men college graduates of our group with that reported for the generality of men graduates indicates that the marriage rate is about the same for both. Our women graduates, however, are more likely to marry than are the generality of college women: 86 percent of the gifted women graduates (all ages) have married as compared with 74 percent of college women in general in the 40–49-year age bracket and 65 percent of those age 50 and over.[17] The women college graduates, although they are less likely to remain spinsters than are college women in general, have a somewhat lower marriage rate than that of the noncollege women in our group. Table 57 gives the proportion of gifted men and women at three educational levels who have married.

TABLE 57

INCIDENCE OF MARRIAGE ACCORDING TO EDUCATIONAL LEVEL

| | Percentages | | | |
| | Men | | Women | |
Education	Single	Are, or have been married	Single	Are, or have been married
College graduates	7.6	92.4	13.8	86.2
College 1 to 4 years	6.4	93.6	4.9	95.1
No college	5.0	95.0	2.1	97.9

Slightly more than one-fifth of those who married have a history of divorce. There are 150 men and 121 women who have been divorced one or more times. These figures represent 21 percent of the men and 22 percent of women who have married. Of these, 32 men (4.1%) and 29 women (4.7%) have been divorced two or more times, and 6 men and 13 women have been divorced three or more times. It is impossible to say how the divorce rate in the gifted group compares with that for the generality of corresponding age since the census data report only current marital status. However, recent estimates of the proportion of marital failures among the general population of the United States place the figure at one-fourth to one-third of the marriages formed.[9] According to the figures for our gifted group the divorce rate to 1955 is somewhat less than that for the generality. However, it is still too soon to say how the ultimate divorce rate will compare with that of the total population since additional divorces can be expected among our subjects.

Divorce is negatively associated with extent of schooling. Those subjects who graduated from college have a far lower divorce rate than do the nongraduates. Of the college graduates who married, only 16 percent have a history of divorce while 32 percent of those who attended college one or more years without graduating and 36 percent of those who did not enter college have a record of one or more marital failures. The trend toward a lower divorce rate for those who completed college agrees with data reported by other investigators for college graduate populations. In a questionnaire survey made in 1950 of the Harvard class of 1926, in which replies were received from 61 percent, the figures indicated that 13 percent of those who had married had been divorced one or more times.[14] An investigation made by Havemann and West[17] of the total U.S. college graduate population showed the proportion divorced at the time of the survey to be 6 percent. However, the latter inquiry did not include a record of marital history and because of the known tendency of college graduates to remarry soon after divorce, this estimate is undoubtedly too low. Table 58 shows, separately for men and women, according to educational level the proportion who (1) are single, (2) are married (no divorce), and (3) have been divorced one or more times (may currently be married).

TABLE 58

MARITAL STATUS BY AMOUNT OF EDUCATION

Marital Status	College Graduates		College 1 to 4 Years		No College		Total Group	
	Men (N = 555)	Women (N = 413)	Men (N = 125)	Women (N = 101)	Men (N = 100)	Women (N = 96)	Men (N = 780)	Women (N = 610)
Single	7.6	13.8	6.4	4.9	5.0	2.1	7.0	10.5
Married, no divorce	77.8	72.4	60.8	70.3	67.0	57.3	73.7	69.7
Divorced 1 or more times	14.6	13.8	32.8	24.8	28.0	40.6	19.3	19.8
% of N ever married who have been divorced	15.8	16.0	35.0	26.0	29.5	41.6	20.7	22.1

Among men, the divorce rate is highest for those who attended college one or more years without graduating, and for women the proportion of divorces is greatest for those who did not enter college. One might speculate that greater restlessness, discontent, or frustration are felt by the gifted persons who do not complete college and that this may bring about greater instability in personal relationships.

The data for women support this theory, at least to the extent that general adjustment ratings are related to both education and divorce. Those women who either did not complete or did not enter college are less often rated satisfactory in mental health and general adjustment. The proportion rated satisfactory at each of these two educational levels is 58 percent, compared with 69 percent of the college graduates. In the case of men the relationship of education to general adjustment rating is quite different. The college graduates and those with one to four years of college have almost exactly the same proportion of satisfactory ratings, 69 and 68 percent, respectively. But the highest proportion of satisfactory ratings (73%) is found for the men who did not enter college! As would be expected, those persons with a history of divorce are much less likely to be rated satisfactory in general adjustment. The proportion of men rated satisfactory is 75 percent for the unbroken marriages and 57 percent for those with a divorce history. The difference is even more marked for women with 71 percent of persons with unbroken marriages rated satisfactory as against 46 percent of those with a record of divorce.

Neither marriage nor divorce rate in this group is correlated with childhood IQ for either men or women. The adult Concept Mastery test scores show practically no difference between the married with no divorce record and those with a history of divorce. The highest CMT scores, however, were made by the single men and the single women! The 34 single men scored about 17 points higher, on the average, than those who had married (with or without divorce). The difference was somewhat less in the case of women, with the single women averaging a little over 8 points higher than those who had married.

If the divorce rate in our group as a whole is high, so also is the rate of remarriage. Of the 150 men who have been divorced, 86 percent remarried at least once. Of the 121 women with a history of divorce, slightly more than two-thirds remarried. When the gifted subjects divorce and remarry they tend to make happy remarriages. In 1950, the subjects were given a shortened form of the 1940 test of marital happiness which has been reported in detail in earlier publications.[36, 38] In the 1950 test the maximum happiness score was 44 points. For both men and women there is virtually no difference in the happiness score between those whose first marriage was intact and those with a history of divorce(s) and remarriage(s). The mean scores by marital history of currently married men and women follow:

	Men (N = 515)		Women (N = 393)	
	Married, no divorce	Divorced, remarried 1 or more times	Married, no divorce	Divorced, remarried 1 or more times
Mean Happiness Score	23.7	25.6	25.5	25.4
S.D.	9.7	8.8	9.8	10.3

Other data collected in the 1950–52 field follow-up throw additional light on the marriages and marital happiness of the gifted. For example, the questionnaire, The Happiness of Your Marriage, included the item *How happy has your marriage been?* The subjects and their spouses were asked to check the appropriate description from a list of seven response choices. The percentage distribution of responses for the gifted and spouses who replied were as follows:

	Gifted Men (N = 515) %	Wives of Gifted Men (N = 327) %	Gifted Women (N = 393) %	Husbands of Gifted Women (N = 242) %
Extraordinarily happy	26.8	35.8	27.7	33.9
Decidedly more happy than average	42.1	33.6	46.1	40.5
Somewhat more happy than average	14.2	17.4	12.7	11.6
Average	10.7	9.5	7.6	7.8
Somewhat less happy than average	3.1	2.8	2.3	2.5
Decidedly less happy than average	2.9	0.6	2.6	2.9
Extremely unhappy	0.2	0.3	1.0	0.8

Various studies[19] of marital happiness for the population in general indicate that about 65 percent of married couples are happy or very happy, about 20 percent just get along (probably equivalent to our rating of average), and about 15 percent are more or less unhappy. In contrast to these figures, better than 85 percent of our subjects and their spouses rate their marriage above average in happiness and only about 6 percent say they are less happy than average (actually only 3.7% of wives of gifted men say they are more or less unhappy). Furthermore, according to the subjects' Supplementary Biographical Data blank 73 percent of gifted men and 70 percent of gifted women consider their marriage to be an aspect of life from which the greatest satisfaction is derived.

A discussion of marriage in the gifted group would not be complete without a description of the men and women they marry. In the matter of age, the gifted subjects follow the general pattern in our culture by choosing wives who are younger and husbands who are older than themselves. The gifted men tend to marry women about two and one-half years younger while the gifted women as a rule choose husbands about three and one-half years older. These, however, are only the averages and there were many exceptions. For example, 8.6 percent of the gifted men chose wives who were two or more years older, including four men whose wives were from 8 to 12 years older. On the other hand, 6 percent of gifted men married women 10 or more years younger than themselves, and in one instance the gifted husband (his second marriage) was 22 years older than his wife. As for the gifted women, close to 10 percent married men who were two to ten years younger, while seven women chose husbands from 20 to 26 years older than themselves.

Investigations agree in showing education to be an important factor in marital selection, so it is not surprising to find that many of our gifted subjects have married college graduates. More than one-half of the husbands and more than two-fifths of the wives graduated from college. The proportion of college graduates who have taken degrees beyond the bachelor's is 24 percent of husbands and 10 percent of wives. For the 514 husbands and 684 wives for whom adequate information on education was available, the percentages with varying amounts of schooling are as follows:

	Husbands %	Wives %
College graduation	55.6	42.4
College 1 to 4 years (no degree)	18.7	26.5
High-school graduation	19.9	27.3
High school 1 to 3 years	3.7	2.5
No formal schooling beyond eighth grade	2.1	1.3

As might be expected in view of their superior education, a large proportion of the husbands are in Group I (professional) and Group II (semiprofessional and higher business occupations) when classified according to the Minnesota Occupational Scale described in Chapter VII. Table 59 gives the percentage distribution by occupational group for the husbands of gifted women. On the distaff side considerably fewer of the wives of gifted men are gainfully employed than are the married gifted women. Whether this is a reflection of the superior economic status of the gifted men or of the greater desire on the part of the gifted women for a career outside the home is a matter for speculation. In any case, only 15.5 percent of the wives of gifted men are employed as compared with 29 percent of the married gifted women. Of the employed wives, 11 percent are in college teaching, research, or the higher professions, 23 percent are in schoolteaching or administration (below college level), and 22 percent are in other professional occupations. About 39 percent are in business and office work, and slightly more than 5 percent are in miscellaneous occupations (e.g., telephone operator, hairdresser, *et cetera*).

TABLE 59

OCCUPATIONAL CLASSIFICATION OF HUSBANDS OF GIFTED WOMEN

Occupational Group	Percentage of Husbands of Gifted Women (N = 487)
I. Professional	34.9
II. Managerial, official, and semiprofessional	39.0
III. Retail business, clerical, skilled crafts, and kindred...	18.7
IV. Agriculture and related occupations	4.3
V. Semiskilled	3.1

Evidence of the generally superior intellectual caliber of the spouses is found in the Concept Mastery test scores. In the field follow-up of 1950–52 the Concept Mastery test, Form T, was given to 690 spouses including 273 husbands and 417 wives. The mean CMT score for the spouses was 95.3 and the standard deviation, 42.7. Although their average score is close to one S.D. lower than the mean score of the gifted subjects, about one-fifth scored above the average for the gifted group. The spouses appear to good advantage when compared to various other groups tested on the CMT. The data for these comparisons are given in Tables 15 and 17 of Chapter V. The mean CMT score of 115 found for the total college graduate group among the spouses is higher than the average of such college graduate populations as the Electronic Engineers and Scientists (Mean score = 94), the applicants to the graduate Public Health Education curriculum (Mean score = 97), and the 75 miscellaneous college graduates (Mean score = 112). For those spouses who have taken graduate degrees, the mean CMT scores range from 149 for those with a Ph.D. to 131 for those with a master's or equivalent degree. In contrast, two groups of advanced graduate students at a leading university made average scores on the CMT of 118 and 119, respectively. The spouses who entered but did not complete college score above the college graduates in the Air Force group (means of 85 and 73, respectively). Moreover, the spouses who did not go beyond high school averaged slightly higher on the CMT than did the total Air Force group, both graduates and nongraduates.

Since in the original survey of 1921–22 the majority of the subjects were chosen from fairly limited geographical areas (chiefly the San Francisco Bay cities and the Los Angeles metropolitan district), some intermarriages among them could be expected. Actually, in ten instances a gifted subject has married a member of the group; however, three of these marriages terminated in divorce. In two cases, the marriage was very brief and there were no children; the third couple were married for 12 years and had two children before being divorced. Among the seven unbroken marriages one couple has 4 children, two couples have 2 children each, one has 1 child, and three couples are childless.

THEIR CHILDREN

By 1955 the gifted group had produced approximately 2,500 children. This number includes both living and deceased offspring and also takes into account the children of those subjects who had died or whose marriages had been terminated by divorce or death of the spouse. Figured on this basis, the average number of children for all subjects who had ever been married was 1.9.

The typical family with children has 2.4 children. The gifted men have slightly larger families with 2.5 children as compared with 2.3 for the gifted women. The two largest families, however, are those of gifted women, one of whom has 8 children and the other 7. There are also six gifted women and three gifted men who have families of 6 children. At the other extreme, among those who have married, 23 percent of the women and 16 percent of the men have no children. Table 60 gives the percentage distribution of family size for all subjects who have ever been married.

The sex ratio among the offspring is 107.8 boys to 100 girls. This represents a slightly greater excess of boys than that reported for the

TABLE 60

NUMBER OF CHILDREN FOR ALL SUBJECTS EVER MARRIED

| | Percentages | | |
Number of Children Per Family	Gifted Men	Gifted Women	Total
5 or more	3.1	2.8	3.0
4	9.7	6.4	8.3
3	19.6	17.5	18.7
2	33.8	30.9	32.5
1	17.6	18.9	18.2
None	16.2	23.4	19.3

total U.S. white population. Dublin gives the sex ratio for the generality as 106 boys born to every 100 girls.[10] Among the offspring are 36 pairs of twins, including one family with three sets and another family with two sets. This is 1.5 percent of the births, which is a slightly greater frequency of twins than that given for the United States as a whole, where during the years 1941–47 there was one set of twins for every 95 births (1.05%).[10]

Of the 2,452 offspring born to 1955, a total of 84 (3.5%) have died and an additional 15 stillbirths have been reported. More than half (56%) of the deaths occurred in the first year of life. Those children who survived the first year died from a number of causes, of which accidents were the most frequent with 11 deaths, and leukemia second with 5 deaths. Age at death ranged from a few hours after birth to 28 years, but all except 7 deaths took place before age 5. Two children died of brain tumor at ages 12 and 14, respectively; 3 were accidentally killed (automobile) at ages 11, 12, and 28; one died of a heart condition at 8 years; and one of bulbar poliomyelitis at 19 years of age.

It is too early to predict the ultimate fertility rate of the gifted group. At the time of last report nearly two-thirds of the gifted women who had married were under 45 years of age. Ten percent were in the 35-to-39-year age bracket and 2 percent were under 35 years of age. The wives of the gifted men were younger, with 71 percent under 45 years of age. About one-fourth of the wives were age 35 to 39 years, and 9 percent were under age 35. Whether the present birth rate of 2.4 children per mother will increase sufficiently to equal the 2.8 children per mother required to maintain the stock remains to be seen.[10] However, in view of the number of child-bearing years remaining to both the gifted women and the wives of the gifted men, a considerable increase in the number of children can be expected.

A comparison of family size with extent of education shows a tendency for the college graduates among the gifted subjects to have somewhat larger families, with the differences according to education being greater for men than for women. The following percentages illustrate the differences in family size at three levels of educational attainment:

	No Children		3 or More Children	
	Gifted Men %	Gifted Women %	Gifted Men %	Gifted Women %
College graduates	14.0	23.0	36.9	28.9
College 1 to 4 years	17.3	21.8	19.2	28.7
No college	19.0	21.7	29.8	21.7

The larger families of the college graduates may be a reflection of their superior economic status. Our data show that those subjects with the larger families have somewhat greater total family incomes. The median total family income in 1954 of couples with no children was $10,462 in comparison with $11,688 for those with 3 or more children. Of the subjects with total family incomes of $25,000 and above, 52 percent had 3 or more children while only about 29 percent of those with incomes below $10,000 had as many as 3 children. However, the relationships are only suggestive and should be interpreted with caution since age also is related both to income and to size of family.

In the field follow-up investigations of 1939–40 and 1951–52 the field workers gave Stanford-Binet tests to all of the offspring of suitable age who could be reached. In addition, a number of subjects who live at a distance have, over the years, brought their children to Stanford University for testing.* Altogether we have tested a total of 1,525 offspring, 786 boys and 739 girls. The distributions of IQ's with means and S.D.'s are given in Table 61.

TABLE 61

STANFORD-BINET IQ's OF OFFSPRING

Binet IQ	Boys	Girls	Total
190–199	1	1	2
180–189	3	7	10
170–179	15	9	24
160–169	22	33	55
150–159	82	66	148
140–149	137	125	262
130–139	188	180	368
120–129	181	168	349
110–119	86	83	169
100–109	53	48	101
90–99	13	11	24
80–89	4	2	6
70–79	1	6	7
Total	786	739	1,525
Mean Score	132.7	132.7	132.7
S.D.	17.2	18.0	17.6

The mean IQ was exactly the same for boys and girls, namely, 132.7 and the S.D.'s differed only slightly in the direction of greater

* As pointed out earlier, the subjects who live outside California and could not be seen by a field worker have been encouraged to call at the research headquarters for a personal interview whenever they might be visiting in the area.

variability among girls. Approximately one-third of the offspring tested at IQ 140 and above while about 2 percent had IQ's below 100. An additional 13 offspring (7 boys and 6 girls) to whom we did not give the Stanford-Binet were known to be mentally defective. Of these, 4 have died, 6 are in schools or institutions for the mentally retarded, and 3 are being cared for at home. Of the total 2,452 offspring, the 13 mentally defective constitute only one-half of one percent, and of the 1,525 offspring whose intellectual status has been determined, they represent only 0.8 percent. In the generality the proportion of the mentally defective (defined as below IQ 70) is very much greater. Of the nearly 3,000 subjects on whom the 1937 revision of the Stanford-Binet test was standardized, 2.6 percent tested below 70. This percentage, however, does not fully represent the incidence of the lower levels of mental ability in the generality age 2 to 18 years since the extreme cases of mental deficiency are not found in the regular school classes.[28]

Some 50 gifted subjects have adopted children. One family has taken four children for adoption; one has adopted 3, and several have 2 adopted children. Although it would have been interesting to have given intelligence tests to all the adopted children, limitations of time and funds made it impractical. However, 18 of these children have been tested and found to have Binet IQ's ranging from about 100 to 146 with 6 of the adopted children testing at IQ 135 or higher.

Although at the present writing a large proportion of our subjects are still under 45 years of age, more than 50 have already reached grandparent status with a total of 115 grandchildren so far reported. One of our gifted women who herself had 5 children is the grandmother of 11 at age 52.

CHAPTER XI

THE FULFILLMENT OF PROMISE

In the past 35 years we have watched the gifted child advance through adolescence and youth into young manhood and womanhood and on into the fuller maturity of mid-life. The follow-up for three and one-half decades has shown that the superior child, with few exceptions, becomes the able adult, superior in nearly every aspect to the generality. But, as in childhood, this superiority is not equally great in all areas.

The superiority of the group is greatest in intellectual ability, in scholastic accomplishment, and in vocational achievements. Physically the gifted subjects continue to be above average as shown in their lower mortality record and in the health ratings. While personal adjustment and emotional stability are more difficult to evaluate, the indications are that the group does not differ greatly from the generality in the extent of personality and adjustment problems as shown by mental breakdowns, suicide, and marital failures. The incidence of such other problems as excessive use of liquor (alcoholism) and homosexuality is below that found in the total population, and the delinquency rate is but a small fraction of that in the generality. Clearly, desirable traits tend to go together. No negative correlations were found between intelligence and size, strength, physical well-being, or emotional stability. Rather, where correlations occur, they tend to be positive.

The Maintenance of Intellectual Ability

But if gifted children are not prone to die young or, as they advance in years, to become invalids or to suffer to any extent from serious personality or behavior difficulties, there remains the question of the degree to which their intellectual superiority is maintained. The evidence on this score is conclusive. Test scores of 1927–28, 1939–40, and 1950–52 showed the majority of the subjects close to the 99th percentile of the generality in mental ability. This is true even of those whose careers have not been particularly notable. It was especially interesting to find that the average Concept Mastery test score in 1950–52 of the subjects who did not go beyond high school was exactly the same as that of a group of candidates for advanced degrees (Ph.D. or M.D.) at a leading university. Of additional interest are the results of a comparison of Concept Mastery test scores of 1939–40 and 1950–52 of the same individuals. The test-retest comparisons showed a reliable gain in the 11-to-12-year interval with increases occurring at all educational and occupational levels, in all grades of ability, and at all ages. The data indicate that not only do the mentally superior hold their own but that they actually increase in intellectual stature as measured by the Concept Mastery test.

Appraisal of Achievement

From a practical and utilitarian point of view the real test of the significance and value of this high degree of mental ability is the use that is made of such gifts. The record points to the conclusion that capacity to achieve far beyond the average can be detected early in life through tests of general intelligence. Such tests do not, however, enable us to predict what direction the achievement will take, and least of all do they tell us what personality factors or what accidents of fortune will affect the fruition of exceptional ability. The appraisal of achievement of our gifted subjects will be concerned with their educational attainments, their vocational records, their contributions to knowledge and culture, and the recognitions that have been won.

The educational record is a distinguished one. More than 85 percent of the group entered college and almost 70 percent graduated. The latter figure is about ten times as high as for a random group of comparable age. Graduation honors and elections to Phi Beta Kappa and Sigma Xi were at least three times as numerous as in the typical senior college class, with better than 35 percent of the graduates winning one or more of these distinctions. Of the college graduates, two-thirds of the men and nearly three-fifths of the women continued for graduate study. The Ph.D. or comparable doctorate was taken by 80 men and 17 women, or about 14 percent of men and 4 percent of women graduates. The proportion of the generality of college graduates of corresponding age who have taken a doctorate is less than 3 percent.

The occupations and occupational status of the men and women of the gifted group have been evaluated separately since the pattern in this regard has been so different. The careers of women are often determined by extraneous circumstances rather than by training, talent, or vocational interest. Whether women choose to work and the occupations they enter are influenced both by their own attitudes and by the attitudes of society toward the role of women. These attitudinal factors also influence the opportunities for employment and for advancement. But in spite of the fact that American women on the average occupy positions of lesser responsibility, opportunity, and remuneration than do men, the gifted women have a number of notable achievements to their credit, some of which have been described in Chapter VII. That 7 women should be listed in *American Men of Science*,[5] 2 in the *Directory of American Scholars*,[6] and 2 in *Who's Who in America*,[42] all before reaching the age of 43, is certainly many times the expectation from a random group of around 700 women. Publications of the gifted women include 5 novels; 5 volumes of poetry and some 70 poems that have appeared in both literary and popular journals; 32 technical, professional, or scholarly books; around 50 short stories; 4 plays; more than 150 essays, critiques, and articles; and more than 200 scientific papers. At least 5 patents have been taken out by gifted women. These figures do not include the writings of reporters and editors, nor a variety of miscellaneous contributions.

Our gifted women in the main, however, are housewives, and many who also work outside the home do so more to relieve the monotony of household duties or to supplement the family income rather than through a desire for a serious career. There are many intangible kinds of accomplishment and success open to the housewife, and it is debatable whether the fact that a majority of gifted women prefer housewifery to more intellectual pursuits represents a net waste of brainpower. Although it is possible by means of rating scales to measure with fair accuracy the achievement of a scientist or a professional or business man, no one has yet devised a way to measure the contribution of a woman who makes her marriage a success, inspires her husband, and sends forth well-trained children into the world.

As for the men, close to three and a half decades after their selection solely on the ability to score in the top one percent of the school population in an intelligence test, we find 86 percent in the two highest occupational categories: I, the professions, and II, the semiprofessions and higher business. Eleven percent are in smaller retail business, clerical, and skilled occupations. Farming and related occupations account for nearly 2 percent and the remaining 1 percent are in semiskilled work. The representation in the two highest groups is many times their proportionate share, with a corresponding shortage of gifted representation in the middle occupational levels. No gifted men are classified in the lowest levels of the occupational hierarchy (service workers and slightly skilled or unskilled laborers), whereas 13 percent of the total urban population are in these categories.

A number of men have made substantial contributions to the physical, biological, and social sciences. These include members of university faculties as well as scientists in various fields who are engaged in research either in industry or in privately endowed or government-sponsored research laboratories.* Listings in *American Men of Science*[5] include 70 gifted men, of whom 39 are in the physical sciences, 22 in the biological sciences, and 9 in the social sciences. These listings are several times as numerous as would be found for unselected college graduates. An even greater distinction has been won by the three men who have been elected to the National Academy of Sciences, one of the highest honors accorded American scientists. Not all the notable achievements have been in the sciences; many examples of distinguished accomplishment are found in nearly all fields of endeavor.

Some idea of the distinction and versatility of the group may be found in biographical listings. In addition to the 70 men listed in *American Men of Science*, 10 others appear in the *Directory of American Scholars,* a companion volume of biographies of persons with notable accomplishment in the humanities.[6] In both of these volumes, listings depend on the amount of attention the individual's work has attracted from others in his field. Listings in *Who's Who in America*,[42] on the other hand, are of persons who, by reasons of outstanding achievement, are subjects of extensive and general interest. The 31 men (about 4%) who appear in *Who's Who* provide striking evidence of the range of talent to be found in this group. Of these, 13 are members of college faculties representing the sciences, arts and humanities; 8 are top-ranking executives in business or industry; and 3 are diplomats. The others in *Who's Who* include a physicist who heads one of the foremost laboratories for research in nuclear energy; an engineer who is a director of research in an aeronautical laboratory; a landscape architect; and a writer and editor. Still others are a farmer who is also a

---

* A detailed study of the vocational correlates and distinguishing characteristics of scientists and nonscientists among the gifted men was made in 1952 under the sponsorship of the Office of Naval Research and has been published in a separate monograph[35] and also appeared in an abbreviated version as an article in the *Scientific American*.[33]

government official serving in the Department of Agriculture; a brigadier general in the United States Army; and a vice-president and director of one of the largest philanthropic foundations.

Several of the college faculty members listed in *Who's Who* hold important administrative positions. These include an internationally known scientist who is provost of a leading university, and a distinguished scholar in the field of literature who is a vice-chancellor at one of the country's largest universities. Another, holding a doctorate in theology, is president of a small denominational college. Others among the college faculty include one of the world's foremost oceanographers and head of a well-known institute of oceanography; a dean of a leading medical school; and a physiologist who is director of an internationally known laboratory and is himself famous both in this country and abroad for his studies in nutrition and related fields.

The background of the eight businessmen listed in *Who's Who* is interesting. Only three prepared for a career in business. These include the president of a food distributing firm of national scope; the controller of one of the leading steel companies in the country; and a vice-president of one of the largest oil companies in the United States. Of the other five business executives, two were trained in the sciences (both hold Ph.D.'s) and one in engineering; the remaining two were both lawyers who specialized in corporation law and are now high-ranking executives. The three men in the diplomatic service are career diplomats in foreign service.

Additional evidence of the productivity and versatility of the men is found in their publications and patents. Nearly 2000 scientific and technical papers and articles and some 60 books and monographs in the sciences, literature, arts, and humanities have been published. Patents granted amount to at least 230. Other writings include 33 novels, about 375 short stories, novelettes, and plays; 60 or more essays, critiques, and sketches; and 265 miscellaneous articles on a variety of subjects. The figures on publications do not include the hundreds of publications by journalists that classify as news stories, editorials, or newspaper columns, nor do they include the hundreds, if not thousands, of radio, television, or motion picture scripts. Neither does the list include the contributions of editors or members of editorial boards of scientific, professional, or literary magazines. There have also been a sizable number of scientific documents reporting studies in connection with government research which are restricted publications. We do not have information on the exact number or content of these.

The foregoing are only a few illustrations of conspicuous achievement and could be multiplied many times. They by no means represent all of the areas or types of success for there is scarcely a line of creditable endeavor in which some member of the group has not achieved outstanding success. There are men in nearly every field who have won national prominence and 8 or 10 who have achieved an international reputation. The latter include several physical scientists, at least one biological scientist, one or two social scientists, two or three members of the United States State Department, and a motion picture director. The majority, though not all so outstanding as those mentioned, have been highly successful vocationally from the standpoint of professional and business accomplishment as measured by responsibility and importance of position, prestige, and income.

There is, however, another side to the picture. There are various criteria of success, but we are concerned here with vocational achievement, and success has been defined as the extent to which the subject has made use of his intellectual ability. This calls for a very high level of accomplishment since the intellectual level is so high and not all have measured up to it vocationally. Although not more than three or possibly four men (again women are not included) could be considered failures in relation to the rest of the group, there are 80 or 90 men whose vocational achievements fall considerably short of the standard set by the group as a whole.

Since the less successful subjects do not differ to any extent in intelligence as measured by tests, it is clear that notable achievement calls for more than a high order of intelligence. After the 1940 follow-up a detailed analysis was made of the life histories of the 150 most successful and 150 least successful men among the gifted subjects in an attempt to identify some of the nonintellectual factors that affect life success. The results of this study indicated that personality factors are extremely important determiners of achievement. The correlation between success and such variables as mental health, emotional stability, and social adjustment is consistently positive rather than negative. In this respect the data run directly counter to the conclusions reached by Lange-Eichbaum in his study of historical geniuses.[24] A number of interesting differences between the two sub-groups were brought out but the four traits on which they differed most widely were "persistence in the accomplishment of ends," "integration toward goals," "self-confidence," and "freedom from inferiority feelings." In the total picture the greatest contrast between the two groups was in all-round emotional and social adjustment, and in drive to achieve. This study is fully reported in *The Gifted Child Grows Up*.[36]

OUTLOOK FOR FUTURE ACHIEVEMENT

The careers of the gifted subjects, now in their mid-forties, are pretty well set in their present courses. In a very few cases, there are no higher rungs on the particular professional or executive ladder they have climbed. But for most of the group, advances to greater levels of achievement and more important roles can be looked for. Lehman[25] has shown that the median age at which positions of leadership are reached has greatly increased in the last 150 years. In field after field the increase has amounted to 8, 10, or even 12 years and numerous positions of high-ranking leadership are most likely to be acquired and retained from fifty to seventy years of age. Lehman has also shown that in nearly all fields of intellectual achievement the most creative period is between thirty and forty-five years. But here Lehman is concerned with *quality* of achievement. Productivity as measured by quantity is often greater after forty than before. And regardless of the merit of one's work, the peak of recognitions, honors, and earned income is usually not reached until the fifties.

On the basis of Lehman's data as well as on the evidence from their own records, the peak of achievement for this group is not yet reached. More than half were still under age 45 in 1955 and there was little evidence of any slackening of pace. Whether the rise in the next 10 years will be as steep as that between 1945 and 1955 is doubtful, principally for the reason that they are so much nearer the top. The group has made tremendous strides in the past ten or fifteen years. This is true in every field and in every walk of life. There is almost no one who has not improved his status, even though he may still be well below the average of the group in terms of realizing his intellectual potential in his vocational accomplishments.

We said some years ago,[36] that only a professed seer would venture a statistical forecast of the future achievements of the group. However, we did venture some predictions on the basis of the data to 1945, among which were the following:

The peak of *recognition* for achievement will come much later, probably not before another fifteen or twenty years have elapsed. Listings in *American Men of Science* may well be doubled by 1960, and listings in *Who's Who* may be trebled or quadrupled by 1970. In the decade 1960 to 1970 there should be several times as many holding positions of high responsibility as in 1945, and several times as many of national or international reputation in their special fields of accomplishment.

These were indeed conservative estimates. Instead of the doubling of listings in *American Men of Science* which was thought might take place by 1960, the number has quadrupled, with 77 names (70 men and 7 women) compared to the 19 men and no women to 1945. The list in *Who's Who in America* has grown from 5 names (all men) to 33 (31 men and 2 women), an increase of more than six times rather than the trebling or quadrupling cautiously predicted for the still-distant 1970.

In 1945 probably not more than a half-dozen had a national reputation, and perhaps one was internationally known. By 1955 several dozen at least have become national figures and 8 or 10 are known internationally. Moreover, the group now includes three men who have been elected to the National Academy of Sciences as compared with only one at the earlier date.

It is hard to say in which fields the greatest advances will take place in the next five or ten years. Business will certainly be one, and law another. The scientists are probably nearer their peak than are the rest of the group but even here there are a number of younger scientists with great promise. Regardless of the degree of productivity yet to be attained, the number of those winning special honors and distinctions will increase. This is true because of the time lag between achievement and recognition. Although *American Men of Science* listings are probably now close to their maximum, at least one and possibly two scientists are so outstanding that eventual election to the National Academy of Sciences can be predicted for them. There will undoubtedly be a considerable increase in the number of *Who's Who* biographies but we hesitate to estimate the ultimate number.

There are, however, a few fields, all dependent on special talent, in which there has been a lack of outstanding accomplishment. These are the fine arts, music and, to a lesser extent, literature. The group has produced no great musical composer* and no great creative artist.

---

* An exception in the case of musical composer should be noted. This is a man of rare creative genius who was not included in the statistics of this report because his intelligence level was not definitely established in childhood. He is several years

Several possessing superior talent in music or art are heading university departments in these fields and have produced some excellent original work, but none seems likely to achieve a truly great piece of creative work. There are a number of competent and highly successful writers among the subjects but not more than three or four with a high order of literary creativity. Perhaps it is not surprising, in view of the relatively small size of our group, that no great creative genius in the arts has appeared, for such genius is indeed rare. In any case these are the only major fields in which the achievement of our group is limited.

Some Comments on Success

Our discussion so far has been concerned with achievement of eminence, professional status, and recognized position in the world of human affairs. But these are goals for which many intelligent men and women do not consciously strive. Greatness of achievement is relative both to the prevailing patterns of culture and the individual's personal philosophy of life; there neither exists nor can be devised a universal yardstick for its measure. The criterion of success used in this study reflects both the present-day social ideology and an avowed bias in favor of achievement that calls for the use of intelligence. It is concerned with vocational accomplishment rather than with the attainment of personal happiness. And the record shows that the gifted subjects, in overwhelming numbers, have fulfilled the promise of their youth in their later life achievements.

There are other criteria of success and other goals and satisfactions in life, however, and in the biographical data blank the gifted men and women have expressed their own opinions on what constitutes life success. The final question in the blank was worded as follows: *From your point of view, what constitutes success in life?* There was a wide range of replies, often overlapping, and frequently a respondent gave more than one definition. The definitions most frequently given fall into five

---

older than anyone included in the group, and when he was a child no satisfactory IQ test had been devised. However, the senior author has followed his development since 1910, when he was 13 years old, and has known him about as intimately as any gifted subject under observation. He is an eminent musician who has produced hundreds of musical compositions, authored two books and scores of articles on musical theory; invented new musical techniques, given recitals throughout the United States and Europe; lectured in leading American universities; founded and edited a musical magazine, and won recognition as an authority on musicology and primitive music. His compositions cover a wide range with respect to type, theme, and technique. Many of his productions have been recorded; several of his orchestral selections are played by leading conductors; and some of his briefer compositions are famous among musicians because of their originality.[36]

categories, each noted by from around 40 to 50 percent of the group (with the exception of category *c*). None of the other definitions of success was mentioned by more than 15 percent, and only two by more than 10 percent of the subjects. The five most frequently mentioned definitions of life success are:

a. Realization of goals, vocational satisfaction, a sense of achievement;

b. A happy marriage and home life, bringing up a family satisfactorily;

c. Adequate income for comfortable living (but this was mentioned by only 20 percent of women);

d. Contributing to knowledge or welfare of mankind; helping others, leaving the world a better place;

e. Peace of mind, well-adjusted personality, adaptability, emotional maturity.

We would agree with the subjects that vocational achievement is not the only—perhaps not even the most important—aspect of life success. To many, the most important achievement in life is happiness, contentment, emotional maturity, integrity. Even failure to rise above the lowest rungs of the occupational ladder does not necessarily mean that success in the truest sense has been trivial. There may have been heroic sacrifices, uncommon judgment in handling the little things of daily life, countless acts of kindness, loyal friendships won, and conscientious discharge of social and civic responsibilities. If we sometimes get discouraged at the rate society progresses, we might take comfort in the thought that some of the small jobs, as well as the larger ones, are being done by gifted people.

REFERENCES CITED

1. ANDERSON, E. E., *et al.* "Wilson College Studies in Psychology: I. A Comparison of the Wechsler-Bellevue, Revised Stanford-Binet, and American Council on Education Tests at the College Level." *Journal of Psychology* (1942), **14**, 317–26.
2. BAYLEY, NANCY, and ODEN, MELITA H. "The Maintenance of Intellectual Ability in Gifted Adults." *Journal of Gerontology* (1955), **10**, 91–107.
3. BURKS, BARBARA S., JENSEN, DORTHA W., and TERMAN, L. M. *Genetic Studies of Genius, III. The Promise of Youth.* Stanford University: Stanford University Press (1930). 508 pages.
4. CADY, V. M. *The Estimation of Juvenile Incorrigibility. Journal of Delinquency* Monograph Series (1923), No. 2. 140 pages.
5. CATTELL, JACQUES (editor). *American Men of Science* (9th edition). *I. The Physical Sciences.* 2,180 pages. *II. The Biological Sciences.* 1,276 pages. *III. The Social & Behavioral Sciences.* 762 pages. Lancaster, Pennsylvania: Science Press (1955).
6. CATTELL, JACQUES (editor). *Directory of American Scholars.* New York: R. R. Bowker Company (1957). 836 pages.
7. CIOCCO, ANTONIO. "Sex Differences in Morbidity and Mortality." *Quarterly Review of Biology* (1940), **15**, 59–73; 192–210.
8. COX, CATHARINE M. *Genetic Studies of Genius, II. The Early Mental Traits of Three Hundred Geniuses.* Stanford University: Stanford University Press (1926). 842 pages.
9. DAVIS, KINGSLEY. "Divorce." Ch. 4, Part V, in Fishbein, Morris, and Burgess, E. W., *Successful Marriage.* Garden City, New York: Doubleday & Company, Inc. (1955), 461–74.
10. DUBLIN, L. I. *Facts of Life.* New York: The Macmillan Company (1951). 461 pages.
11. DUBLIN, L. I., and BUNZEL, B. *To Be Or Not To Be.* New York: Harrison Smith and Robert Haas (1933). 443 pages.
12. DUBLIN, L. I., and LOTKA, A. J. *Length of Life.* New York: The Ronald Press Company (1936). 400 pages.
13. DUBLIN, L. I., LOTKA, A. J., and SPIEGELMAN, MORTIMER. *Length of Life.* New York: The Ronald Press Company (1949). 379 pages.
14. DUBOIS, CORNELIUS, and MURPHY, J. V. *Life and Opinions of a College Class* (Harvard 1926). Cambridge: Harvard University Press (1951). 98 pages.
15. GOLDHAMER, H., and MARSHALL, A. *Psychosis and Civilization.* Glencoe, Illinois: The Free Press (1953). 126 pages.
16. GOODENOUGH, FLORENCE L., and ANDERSON, J. E. *Experimental Child Study.* New York: Century Company (1931). 546 pages.

17. HAVEMANN, ERNEST, and WEST, PATRICIA SALTER. *They Went to College.* New York: Harcourt, Brace and Company (1952). 227 pages.
18. HENRY, A. F., and SHORT, J. F., JR. *Suicide and Homicide.* Glencoe, Illinois: The Free Press (1954). 214 pages.
19. HIRNING, J. L., and HIRNING, ALMA L. *Marriage Adjustment.* New York: American Book Company (1956). 456 pages.
20. JELLINEK, E. M., and KELLER, MARK. "Rates of Alcoholism in the United States of America, 1940–1948." *Quarterly Journal of Studies on Alcohol* (1952), **13**, 49–59.
21. KINSEY, A. C., POMEROY, W. B., and MARTIN, C. E. *Sexual Behavior in the Human Male.* Philadelphia: W. B. Saunders Company (1948). 804 pages.
22. KINSEY, A. C., POMEROY, W. B., MARTIN, C. E., and GEBHARD, P. H. *Sexual Behavior in the Human Female.* Philadelphia: W. B. Saunders Company (1953). 842 pages.
23. LANDIS, CARNEY, and PAGE, J. D. *Modern Society and Mental Disease.* New York: Farrar and Rinehart (1938). 190 pages.
24. LANGE-EICHBAUM, WILHELM. *The Problem of Genius.* New York: The Macmillan Company (1932). 187 pages.
25. LEHMAN, H. C. *Age and Achievement.* Pub. for American Philosophical Society by Princeton University Press (1953). 359 pages.
26. MALZBERG, BENJAMIN. "The Expectation of Mental Disease in New York State, 1920, 1930, and 1940," in American Psychopathological Association, *Trends of Mental Disease.* New York: King's Crown Press (1945), 42–55.
27. McNEMAR, QUINN, and TERMAN, L. M. "Sex Differences in Variational Tendency." *Genetic Psychology Monographs* (1936), **18**, 66 pages.
28. MERRILL, MAUD A. "The Significance of IQ's on the Revised Stanford-Binet Scales." *Journal of Educational Psychology* (1938), **29**, 641–51.
29. RAUBENHEIMER, A. S. "An Experimental Study of Some Behavior Traits of the Potentially Delinquent Boy." *Psychological Monographs* (1925), **34**, No. 159. 107 pages.
30. SEMELMAN, BARBARA BLUM. *A Study of Suicide in San Francisco, 1943–45.* Unpublished master's thesis, Stanford University, 1948.
31. STRONG, E. K. *Vocational Interests of Men and Women.* Stanford University: Stanford University Press (1943). 746 pages.
32. SWARD, KEITH. "Age and Mental Ability in Superior Men." *American Journal of Psychology* (1945), **58**, 443–79.
33. TERMAN, L. M. "Are Scientists Different?" *Scientific American* (1955), **192**, 25–29.
34. TERMAN, L. M. "Genius and Stupidity; a Study of Some of the Intellectual Processes of Seven 'Bright' and Seven 'Stupid' Boys." *Pedagogical Seminary* (1906), **13**, 307–73.

35. TERMAN, L. M. "Scientists and Nonscientists in a Group of 800 Gifted Men." *Psychological Monographs* (1954), **68**, No. 7 (Whole No. 378), 44 pages.

36. TERMAN, L. M., and ODEN, MELITA H. *Genetic Studies of Genius, IV. The Gifted Child Grows Up.* Stanford University: Stanford University Press (1947). 448 pages.

37. TERMAN, L. M., SUMPTION, M. R., and NORRIS, DOROTHY. "Special Education for the Gifted Child." *Yearbook of the National Society for the Study of Education* (1950), **49 (II)**, 259–80.

38. TERMAN, L. M., and WALLIN, PAUL. "The Validity of Marriage Prediction and Marital Adjustment Tests." *American Sociological Review* (1949), **14**, 497–504.

39. TERMAN, L. M., *et al. Genetic Studies of Genius, I. Mental and Physical Traits of a Thousand Gifted Children.* Stanford University: Stanford University Press (1925). 648 pages.

40. U.S. BUREAU OF THE CENSUS. Department of Commerce. *Current Population Reports* (October 1955), Series P60, No. 19. Washington: United States Government Printing Office (1955). 24 pages.

41. U.S. BUREAU OF THE CENSUS. Department of Commerce. *Statistical Abstract of the United States.* Washington: United States Government Printing Office (1957). 1,045 pages.

42. *Who's Who in America* (30th edition). Chicago: A. N. Marquis Company (1958). 3,388 pages.

43. WOLFLE, DAEL. *America's Resources of Specialized Talent.* Report of the Commission on Human Resources and Advanced Training. New York: Harper & Brothers (1954). 332 pages.

44. WOODBURY, R. M. "The Relation Between Breast Feeding and Artificial Feeding and Infant Mortality." *American Journal of Hygiene* (1922), 668–87.

45. *Yearbook of American Churches.* New York: National Council of the Churches of Christ in the U.S.A. (1951). 272 pages.

APPENDIX

APPENDIX

The information blanks used in the 1950–52 field follow-up and in the 1955 survey by mail were 8½ by 11 inches in size. They are reproduced here in small type. The list that follows includes all the blanks used at these dates except two: (1) a 4-page blank titled Information About Child which called for developmental data on those offspring of the gifted subjects who had been given a Stanford-Binet test; and (2) the Concept Mastery test.

Blanks reproduced:
    General Information (1950)
    Supplementary Biographical Data
    Data on Rate of Reproduction
    The Happiness of Your Marriage
    Report of Field Worker
    Information Blank (1955)

Gifted Children Follow-Up
Stanford University 1950
Date of filling out this blank...

GENERAL INFORMATION

Full name .... Birthdate.. Age...................

(Married women include maiden name)

Address .... Telephone.................................

Name and address of relative or friend through whom you could be reached if your address should change ....

....

1. Education: Circle highest grade completed. High school 1 2 3 4   College 1 2 3 4   Postgraduate 1 2 3 4

   Since 1940: College degrees.................................. Date(s).................................. College attended..........................

   Special courses taken (extension, business, technical, professional, etc.) ....

   ....

2. Occupation and earned income (for income report annual salary *before* income tax deductions). If self-employed (doctor, lawyer, business owner, etc.) give equivalent of salary, i.e., income less operating expenses.

Year	Profession, job, or position	Exact nature of work	Earned income per year
1946			
1947			
1948			
1949			

   Approximate current income from other sources (investments, trust funds, other assets) ....

3. Marital status (check) single........ ; married........ ; widowed........ ; separated........ ; divorced........

   Date of marriage.......................... Your age at marriage: Years............................ Months.............................

   If this or a previous marriage ended in divorce, give date(s) ....

4. About your spouse: Name (maiden name of wife).. Age of spouse at

   marriage: Years........................ Months........................ Highest grade or college year of spouse's schooling.........................

   Diplomas or degrees received.. What school or college? ....

   Present occupation .... Present annual earned income............................

   His (her) avocational interests or hobbies...

   ....

   Occupation of his (her) father.. His (her) mother...

   Other information regarding husband (wife) that you think would be of interest ....

   ....

5. Offspring:

Name	Sex	Date of birth	School grade	If not living, age at death	Cause of death

5. Your general health since 1945:

a) Physical condition has been: (check) Very good..........; good..........; fair..........; poor..........; very poor..........

b) Illnesses, accidents, or surgical operations in recent years ....

....

Aftereffects ....

c) Has there been any tendency toward nervousness, worry, special anxieties, or nervous breakdown.................. Date

and nature of such difficulties ....

....

....

How handled (psychiatric, psychoanalytic, or medical help, hospitalization, etc.) Give details ....

....

Present condition (free from difficulty, improved, no change, worse, etc.) ....

....

d) Use of liquor (check the statement below that most nearly describes you)

..........I never take a drink, or only on rare occasions.

..........I am a moderate drinker. I have seldom or never been intoxicated.

..........I am a fairly heavy drinker; I drink to excess rather frequently but do not feel that it has interfered seriously with my work or relationships with others.

..........Alcohol is a serious problem. I am frequently drunk and attempts to stop drinking have been unsuccessful.

If alcohol has been a serious problem, what steps have been taken? ....

....

7. Have you ever been arrested? (Do not include *minor* traffic violations)...................... If so, give facts regarding each

instance, including date, charge made, and disposition of case..

....

8. List any members of your family (parents, grandparents, brothers, sisters) who have died since 1940.

Relationship	Date of death	Age at death	Cause of death

9. (a) Occupation of father, if living .... Special accomplishments, activities, honors,

or misfortunes of father in last few years ....

....

(b) Occupation of mother, if living .... Special accomplishments, activities, honors,

or misfortunes of mother in last few years ....

....

10. Your brothers and sisters. Mark with cross (X) any half brothers or sisters. If deceased, mark D after name and give age at death.

Name	Age	Amount of education	Occupation	Married Yes or no	No. of children Living	No. of children Deceased

Have any brothers or sisters been divorced?.............. If so, which (indicate if divorced more than once).......................

Special accomplishments, activities, honors, or misfortunes of brothers or sisters in recent years ....

11. (a) List your avocational interests or hobbies of recent years (e.g., sports, music, art, writing, collections, gardening, woodwork, etc.) Underline each activity *once* to show moderate interest, *twice* to show very great interest.

Special instruction in any of above interests (amount and kind) ....

(b) What kind of reading do you prefer? (fiction, biography, poetry, etc.) ....

Give illustrations from your reading of the last year ....

What magazines do you read fairly regularly? ....

12. How regularly do you vote? (check for each kind of election)

National election: Always........ Usually........ Occasionally........ Rarely........ Never........

State election: Always........ Usually........ Occasionally........ Rarely........ Never........

Local election: Always........ Usually........ Occasionally........ Rarely........ Never........

13. On national issues which of the political parties most nearly represents your leanings? (check) Democrat........ Republican........ Socialist........ Communist........ Other (specify)...................................

Have you held political office? (specify) ....

Other political activities ....

14. Rate yourself on the following scale as regards your political and economic viewpoint (Indicate by cross (X) on the line)

—|—————————————|—————————|—————————|————————————|————

| Extremely radical | Tend to be radical | Average | Tend to be conservative | Extremely conservative |

5. List memberships in clubs or organizations and offices held (e.g., labor unions, business or professional organizations, social clubs, service organizations, etc.) ....

....

....

6. Record of your service activities (such as scout work, welfare, religious or church work, participation in community and civic affairs, P.T.A., etc.) Please do not be modest—include offices held, etc. ....

....

....

7. Any special honors, citations, awards, election to honor societies, "Who's Who" listings, etc? (specify) ....

....

8. List your publications since 1940, if any. Give title, date, publisher, and type of materials (e.g., poem, short story, musical composition, plays, scientific or critical articles, etc.) If space is insufficient, attach further sheets listing such publications. ....

....

....

9. List other creative work accomplished (e.g., architectural, engineering, inventive, scientific, artistic, dramatic). Note any special recognition ....

....

....

20. Give any other significant information regarding yourself or your family which has not been covered in this questionnaire, e.g. (a) marriage of children and birth of grandchildren; (b) any special good fortune, accomplishments, or change of status; (c) any misfortunes or disappointments that have seriously affected your life, etc. (If space is inadequate, answer on additional sheet).

....

....

....

....

....

....

....

....

....

....

....

....

....

....

....

....

....

Gifted Children Follow-Up
Stanford University, 1950–51

Date of filling blank.. Age............

Name ....
(Married women include maiden name)

(Name to be cut out on this line)

Code number..........................

SUPPLEMENTARY BIOGRAPHICAL DATA

Directions. The purpose of this questionnaire is to obtain certain kinds of biographical information that will throw light on your personality development and on the factors that may have helped or hindered you in achieving your life goals. The information called for is intended to supplement, clarify, and perhaps, in some cases, correct the information I have collected from all sources over the many years I have known you. Your point of view regarding certain aspects of your early life on which your parents and teachers have reported should be of special value.

Such information as you are willing to give me is, of course, for my confidential records. As soon as the blank has reached me your name at the top of this page will be cut out and a code number will be substituted for it so that no assistant who tabulates the data for statistical treatment will know the identity of any respondent.

Experience has shown that most persons can fill out the blank in an hour or less. If the time allowed does not permit you to finish, you will be given an extra blank to take home with you which you can return to me by mail with whatever additions or amplifications you wish to make.

LEWIS M. TERMAN

1. Following are several aspects of parent-child relationships. Each trait or attitude is represented by a straight line, the two ends of the line being the extremes. Place a cross (x) on the line where you think it should go to describe your relationships with your parents correctly. The cross does *not* have to be placed at the small vertical bars; put it wherever you think it should go.

Try to view your childhood and youth objectively. It should be remembered that frequently either end of the scale might be considered favorable from an independent observer's standpoint; hence the objective rating should be shown even though such a rating might not seem complimentary from your point of view.

a) To what extent did you admire and want to emulate your parents?

FATHER

Extremely   A good deal   Moderately   Slightly   Not at all

MOTHER

Extremely   A good deal   Moderately   Slightly   Not at all

b) To what extent did you feel rebellious toward your parents?

FATHER

Extremely   A good deal   Moderately   Slightly   Not at all

MOTHER

Extremely   A good deal   Moderately   Slightly   Not at all

c) To what extent did your parents encourage your efforts toward initiative and independence?

FATHER

Extremely   A good deal   Moderately   Slightly   Not at all

MOTHER

Extremely   A good deal   Moderately   Slightly   Not at all

d) To what extent did your parents resist your efforts to achieve normal independence from them?

FATHER

Extremely   A good deal   Moderately   Slightly   Not at all

MOTHER

Extremely   A good deal   Moderately   Slightly   Not at all *e*) To what extent did you feel rejected or that your parents' feeling toward you was unsympathetic?

FATHER

MOTHER

|———|———|———|———|
Extremely   A good deal   Moderately   Slightly   Not at all

|———|———|———|———|
Extremely   A good deal   Moderately   Slightly   Not at all

f) To what extent was there a feeling of deep affection and understanding between you and your parents?

FATHER

MOTHER

|———|———|———|———|
Extremely   A good deal   Moderately   Slightly   Not at all

|———|———|———|———|
Extremely   A good deal   Moderately   Slightly   Not at all

g) How solicitous were your parents about you; that is, inclined to "anxious" affection, overprotection, and planning for you?

FATHER

MOTHER

|———|———|———|———|
Extremely   A good deal   Moderately   Slightly   Not at all

|———|———|———|———|
Extremely   A good deal   Moderately   Slightly   Not at all

2. To show what your parents were like in their life as a whole, *entirely apart from their relationships toward you or their other children*, please rate them on the following traits:

a) How self-confident?

FATHER

MOTHER

|———|———|———|———|
Extremely   More than average   Average   Less than average   Markedly lacking

|———|———|———|———|
Extremely   More than average   Average   Less than average   Markedly lacking

b) How helpful?

FATHER

MOTHER

|———|———|———|———|
Extremely   More than average   Average   Less than average   Markedly lacking

|———|———|———|———|
Extremely   More than average   Average   Less than average   Markedly lacking

c) How domineering?

FATHER

MOTHER

|———|———|———|———|
Extremely   More than average   Average   Less than average   Markedly lacking

|———|———|———|———|
Extremely   More than average   Average   Less than average   Markedly lacking

d) How friendly?

FATHER

MOTHER

|———|———|———|———|
Extremely   More than average   Average   Less than average   Markedly lacking

|———|———|———|———|
Extremely   More than average   Average   Less than average   Markedly lacking

e) How intelligent?

FATHER

MOTHER

|———|———|———|———|
Extremely   More than average   Average   Less than average   Markedly lacking

|———|———|———|———|
Extremely   More than average   Average   Less than average   Markedly lacking

3. As an adult, do you find yourself becoming more like your father or mother than you were in adolescence? (check) Father............ Mother............. In what ways do you seem to be changing toward or away from your parents' personalities or attitudes?...

....

4. Additional comments regarding your parents or your relationships with them..

....

5. Relationships within family:

   a) Was there either unusually close attachment or marked rivalry or jealousy between you and any of your brothers or sisters? (check) Attachment............; Rivalry............; If present, between whom?.................................

   Effect on you...

   ....

   ....

   b) Was there marked friction between any of the members of your family?.................. If so, which?.................

   Effect on you...

   ....

   ....

6. Socio-economic factors:

   a) During your childhood and youth was there much financial difficulty in your home? Finances were: (check) Very limited............; Limited............; Adequate............; Abundant............

   b) As regards social position, how did your parents compare with the parents of your schoolmates? (check) Much above average............; Somewhat above average............; About average............; Somewhat below average............; Much below average............

   c) In your elementary and high-school days how successful vocationally did you consider your father? (check) Outstanding............; More successful than average............; About average............; Less successful than average............; Quite unsuccessful............

   d) Comments: ....

   ....

   ....

   ....

7. Other influences:

   a) What persons other than your immediate family have markedly influenced your personality and development? Give age at time and describe..

   ....

   b) Has any book or philosophy had a profound influence on you? Give age at time and describe.................

   ....

8. Your attitude toward religion:

   a) Religious training in childhood and youth: (check) Very strict.........; Considerable.........; Little.........; None.........

   b) As an adult, to what extent are you religiously inclined? (check) Strongly..............; Moderately..............; Little ....; Not at all..............

   c) What is your religious affiliation, if any...

   d) Comments: ....

9. Physical factors:

   a) Characterize the general level of your health during the earlier periods of your life as follows: (check)

      Childhood: Very good............ Good............ Fair............ Poor............ Very poor............

      Adolescence: Very good............ Good............ Fair............ Poor............ Very poor............

   b) Did such factors as physical size, appearance, strength, agility, or physical handicap affect your personality, and if so, in what way?...

   ....

   ....

c) What important illnesses or accidents have influenced your personality or attitudes? Explain...................................

....

....

d) How would you rate yourself as to amount of physical energy *in recent years*? (check) Extremely energetic...........;
More energetic and active than average............; Average............; Less energetic and active than average............;
Markedly lacking in energy, fatigue very easily............

e) Comments: ....

....

....

10. Emotional and social factors:

a) As compared to your friends, to what extent were you interested in **succeeding** at the following? (Indicate by a
cross [x] on the line.)

1) Competitive sports

Before age 12	Age 12–20	Since age 20
Ex-tremely A good deal Moder-ately Slightly Not at all	Ex-tremely A good deal Moder-ately Slightly Not at all	Ex-tremely A good deal Moder-ately Slightly Not at all

2) Being a leader

Before age 12	Age 12–20	Since age 20
Ex-tremely A good deal Moder-ately Slightly Not at all	Ex-tremely A good deal Moder-ately Slightly Not at all	Ex-tremely A good deal Moder-ately Slightly Not at all

3) Having friends

Before age 12	Age 12–20	Since age 20
Ex-tremely A good deal Moder-ately Slightly Not at all	Ex-tremely A good deal Moder-ately Slightly Not at all	Ex-tremely A good deal Moder-ately Slightly Not at all

4) Making money

Before age 12	Age 12–20	Since age 20
Ex-tremely A good deal Moder-ately Slightly Not at all	Ex-tremely A good deal Moder-ately Slightly Not at all	Ex-tremely A good deal Moder-ately Slightly Not at all

5) Being a social success

Before age 12	Age 12–20	Since age 20
Ex-tremely A good deal Moder-ately Slightly Not at all	Ex-tremely A good deal Moder-ately Slightly Not at all	Ex-tremely A good deal Moder-ately Slightly Not at all

6) Schoolwork

Before age 12	Age 12–20	Since age 20
Ex-tremely A good deal Moder-ately Slightly Not at all	Ex-tremely A good deal Moder-ately Slightly Not at all	Ex-tremely A good deal Moder-ately Slightly Not at all

b) Did you as a child or later feel that you were "different" from your classmates or associates?............ If so, in what
way did you feel different? ....

Effect on you ....

....

c) In your childhood or adolescence how easy or difficult was it for you to enter into the social and other activities
of your classmates? (check) Had difficulty in making friends and being accepted............; Average in this respect
............; Very adept socially............

Comments: ....

....

....

d) Describe any circumstances, personal influences, or events that contributed to your social adjustment or lack of social adjustment in childhood and youth..

....

....

....

e) Either in childhood or later, have there been any major problems or marked difficulties related to sex?..............
Give age at time and nature of difficulty...

....

....

....

11. Education:

a) Did your parents encourage you to forge ahead in school, were you allowed to go your own pace, or were you held back? (underline)

b) What was the attitude of your parents toward your schoolwork? (check) Demanded high marks............; Encouraged them............; Took them for granted............; Showed little concern............; Were not interested............
Comments: ....

....

c) Did your parents encourage you to go to college?............ Explain: ....

....

d) Did you have as much schooling as you wanted?............ If not, explain..

....

e) Comments: ....

....

....

12. Vocation:

a) What vocation did your parents think you should plan for? ....

....

b) At about what age did you start thinking seriously about your lifework and what vocation(s) appealed to you most at that time?...

....

c) Was there any conflict with your parents regarding your choice of career? Were you greatly influenced by their desires in your choice? Give details...

....

d) Other circumstances that influenced your final choice..

....

e) Which of the following best describes your feeling about your present vocation? (check) Deep satisfaction and interest............; Fairly content............; No serious discontent, but do not find it particularly interesting or satisfying............; Discontented but will probably stick it out............; Strongly dislike and hope to change............

f) If dissatisfied with your vocation, what vocation or vocations do you think would have suited you better?..............

....

g) Comments : ....

....

....

13. Overview :

    a) On the whole, how well do you think you have lived up to your intellectual abilities? (Don't limit your answer to economic or vocational success only.)

        (Check) Fully............; Reasonably well............; Considerably short of it............; Far short of it............; Consider my life largely a failure............; A total failure............

    b) Does your life offer satisfactory outlets for your mental capabilities?............ If not, explain...............................

....

    c) Factors which have *contributed* to your life accomplishment to date : (Check once all of the following that have had a definitely helpful effect; double-check those that have been most helpful.)

        (1) Superior mental ability........; (2) Adequate education............; (3) Good social adjustment............; (4) Good personality........; (5) Good mental stability........; (6) Persistence in working toward a goal........; (7) Good habits of work........; (8) Excellent health........; (9) Lucky "chance" factors (specify)..;

        (10) Helpful factors related to other people (e.g., spouse, children, friends, employer, etc.). Explain.....................

....

        (11) Comments : ....

....

    d) Factors which have *hindered* life accomplishment to date : (check once each thing that has had a definitely hindering effect; double-check those that have hindered most).

        (1) Inadequate mental ability........; (2) Inadequate education........; (3) Poor social adjustment........; (4) Poor personality........; (5) Mental instability........; (6) Lack of persistence in working toward a goal........; (7) Poor work habits........; (8) Poor health........; (9) Unlucky "chance" factors (specify) ....

        (10) Hindering factors related to other people (e.g., spouse, children, friends, employer, etc.). Explain.....................

....

        (11) Comments : ....

....

    e) From what aspects of your life do you derive the greatest satisfaction? (Check once those you regard as important, double-check those most important.)

        Your work itself........; Recognition for your accomplishments........; Your income........; Your avocational activities or hobbies........; Your marriage........; Your children........; Religion........; Social contacts........; Community service activities........; Other..

....

    f) From your point of view, what constitutes "success" in life? ....

....

....

14. Self-ratings on personality traits:

Instructions: Please rate yourself on all of the following traits. For each trait place a cross (x) on the line at the place you think it should go to describe you correctly *as an adult.*

Try to view yourself objectively. Don't hesitate to rate yourself toward the extreme if that is where you belong. The extremes do not necessarily represent clear-cut faults or virtues. Try not to think of the traits in terms of "good" or "bad."

a) How happy is your temperament?

I have an extraordinarily happy and joyous temperament	Have a more happy disposition than the average person	Average in this respect	Usually less happy than others would be in the same circumstances	Am strongly inclined to look on the dark side of things and to be unhappy or discontented

b) Are you moody?

I don't have moods, I always feel about the same	My moods are relatively permanent	Average in this respect	My moods are rather changeable	My moods are decidedly changeable; alternate between extreme joy and extreme sadness

c) How impulsive are you?

Am extremely cautious; I do nothing without considering it from every angle	Am more deliberate and less impulsive than the average	Average in this respect	Rather impulsive; I frequently get into awkward situations because of impulsive action	Extremely impulsive, always rushing headlong into things

d) How self-confident are you?

Extremely self-confident; things that would cause tension and anxiety in most people never worry me	Nearly always confident of myself, more so than the average person	Average in this respect	Less self-confident than the average; am inclined to "borrow trouble"	Extremely lacking in self-confidence; suffer greatly from anxiety and apprehension

e) How emotional are you?

Am extremely *un*emotional, even in situations which arouse strong emotions in others	I tend to be unresponsive to situations of an emotional nature	Average in this respect	Have a tendency to become over-emotional on occasion	Am over-emotional to an extreme degree

f) To what extent do you conform to authority and the conventions?

No tendency whatever to resent or criticize authority, either personal or in the form of conventions	Less resentful than the average	Average in this respect	Am more rebellious than the average	Am inclined to be extremely antagonistic toward authority and conventions

g) In general, how easy are you to get on with?

Extremely good natured, never irritable, always avoid every kind of quarrel or altercation	Am easier to get on with than the average person	Average in this respect	Am harder to get on with than the average person	Am rather inclined to be irritable, quarrelsome, or resentful at slight provocations

h) How much do you enjoy social contacts?

Am socially minded to an extreme; I like nearly everyone and prefer to be with people most of the time	More socially minded than the average	Average in this respect	Less socially minded than the average	Am definitely unsocial; prefer to work and play alone; refuse to be drawn into group activities when I can possibly avoid them

i) How persistent are you in the accomplishment of your ends?

I won't give up; I persevere in the face of every difficulty	Am more persistent than the average person	Average in this respect	Less persistent than the average person	Very easily deterred by obstacles; give up in the face of even trivial difficulties

j) Do you have a program with definite purposes in terms of which you apportion your time and energy?

My life is completely integrated toward a definite goal	I have a well-established plan for my life and usually keep to it	Average in this respect	Am inclined to drift and to be satisfied with just "getting by"	Drift entirely; no definite life plan; leave everything to chance

k) How sensitive are your feelings?

Extremely lacking in sensitiveness; thick-skinned; almost impossible to hurt my feelings	Less sensitive than the average	Average in this respect	More sensitive than the average	Extremely sensitive and thin-skinned. Many things hurt me that others would not notice

l) To what extent have you suffered from feelings of inferiority?

Have rarely or never been conscious of such feelings; I have a feeling of adequacy which almost never deserts me	Have probably suffered less from this cause than the average person	Probably average in this respect	Have probably suffered more from this cause than the average person	Inferiority feelings have been the bane of my life; have suffered agonies from them and still do

15. Supplemental data: Will you please add below any further data or explanations that will contribute to an understanding of your life to date.

....

....

....

....

....

....

....

....

....

....

....

....

....

....

....

....

....

....

....

Gifted Children Follow-up
Stanford University, 1950–51

Name of gifted subject ....
                                                                    (Gifted women include maiden name)

                                                                    (Name to be cut out on this line)

Gifted subject's age............................ Sex (M or F)................................ Date of filling out blank..

Age of spouse..............................   Code number of blank..

DATA ON RATE OF REPRODUCTION

Explanation. The purpose of this questionnaire is to obtain information on factors which may have influenced the rate of reproduction in the gifted group to the present time. Although numerous studies have been made of the number of children born to college graduates and certain other groups, no serious investigation has been made of the *factors which have influenced* fecundity in a large group of intellectually superior persons. Such data on this group would be of great interest to sociologists, psychologists, and students of population trends.

It will be noted that the questionnaire is divided into two parts:

PART I—FOR PERSONS WHO HAVE NEVER BEEN MARRIED

PART II—FOR PERSONS WHO ARE OR HAVE BEEN MARRIED

The data will be used for statistical purposes only, and it is hoped that returns can be obtained from as nearly as possible 100 percent of the group. The blanks will be kept in a separate confidential file. As soon as each blank is returned the name on it will be cut out and replaced by a code number, so that no statistical assistant who works on the material will know the identity of any respondent. The data will be punched on cards, together with other special items of information obtained in the years since the group was first located. The cards will then be run through a card-sorting machine in order to show the relationship between fecundity and each variable or group of variables.

LEWIS M. TERMAN

PART I—FOR SUBJECTS WHO HAVE NEVER BEEN MARRIED

1. How do you feel now about not having married? (Check one of following):

No regret ....; Mild regret ....; Considerable regret ....; Strong regret ....

2. Reasons for not marrying:

(Below are listed numerous possible reasons, each preceded by a dotted line. On the dotted line check once ($\sqrt{}$) each reason that influenced you in any degree. Double check ($\sqrt{}\sqrt{}$) the more important reasons in your case. Triple check ($\sqrt{}\sqrt{}\sqrt{}$) the one most important reason)

(1) .... Feared marriage would interfere with my career

(2) .... Disappointment in love (unhappy love affair, etc.)

(3) .... Responsibility for support of parents or other relatives

(4) .... Poor physical or mental health

(5) .... Bad heredity in my ancestry

(6) .... Little or no interest in physical aspects of sex

(7) .... Positive dislike of physical aspects of sex

(8) .... Preference for companionship of my own sex

(9) .... Feelings of antagonism toward the opposite sex

(10) .... Parental opposition to my getting married

(11) .... Strong attachment between me and my father

(12) .... Strong attachment between me and my mother

(13) .... Marital unhappiness of my parents

(14) .... Just never found the right person

(15) .... Other reasons (specify) ....

PART II—FOR SUBJECTS WHO HAVE MARRIED

1. Number of offspring:

   (a) How many children have been born to you (live births)? ....

   (b) Number of children deceased (including still births)...

   (c) Number of wife's miscarriages (including both spontaneous and induced) ....

2. Is the number of children born to you the number you originally wanted or planned to have? Yes ....; No ....

   If answer is NO, how many did you want or plan for?..

3. If life could be lived over, how many children would you try for? ....

4. Menopause status of wife (check): Menopause not yet started ....; Has begun but is not over ....; Is apparently

   over ....

5. Duration of marriage (if more than one, give data separately for each):

	Your age at marriage (Yrs. & mos.)	Your spouses's age then (Yrs. & mos.)	If marriage ended, indicate how. (Death of spouse, divorce, or separation)	If marriage ended, give date (Yr. & mo.)
First marriage				
Second marriage				
Third marriage				

6. Give for each marriage the approximate number of months you and your spouse were away from each other for any reason, such as war, long trips, impending break-up of the marriage, etc. Do not count any single separation that lasted less than six months.

   First marriage .... ; Second .... ; Third ....

7. Extent to which you and/or your spouse have practiced birth control (check):

   Never ....; Rarely and only for brief periods ....; Regularly for one or two years only ....; Regularly for

   more than two years ....; Approximate number of years during which birth control methods were regularly used

   (that is, total of all such periods in your married life, counting all marriages if there were more than one): ....

8. If birth control methods were never used, or only for very brief periods, give reason (or reasons) why they were not.

   ....

9. Reasons for any birth control practised for as much as one year (check once each reason that operated in any degree, double check all important reasons, and triple check the one most important reason) :

To space pregnancies at appropriate intervals .... ; Inadequate income .... ; Housing difficulties .... ; Husband's

health .... ; Wife's health .... ; Wife's preference for career .... ; Necessary for wife to work in order to supple-

ment husband's income .... ; Wife's dislike of pregnancy or fear of childbirth .... ; Uncertainty due to war or

threat of war .... ; Bad heredity on one or both sides .... ; Unhappiness of the marriage .... ; Husband no de-

sire for children .... ; Wife no desire for children .... ; Other reasons (specify) ....

....

10. Did any pregnancies occur because of failure of birth control methods that were being used? .... ; If so, how many?

....

11. Has it happened that a child was born to you after you had decided that your family was as large as you wanted?

Yes .... ; No .... If answer is Yes, how may times did this occur? ....

12. Has it happened that over a long period of time, when you and your spouse neither practised abstinence nor used

any birth control measures, no pregnancy resulted? Yes .... ; No .... If answer is yes, indicate the approximate

date (or dates) during which conception failed to occur:

From .... to .... ; From .... to .... ; From .... to ....

What do you think was the reason that no pregnancy occurred during such periods? ....

....

13. Have you or your spouse ever consulted a doctor to find out why pregnancy failed to occur? Husband has .... ;

Wife has .... ; Neither .... What was the doctor's opinion about failure of pregnancy to occur?

His opinion about husband ....

His opinion about wife ....

14. (a) How many siblings (i.e., brothers and sisters) have you had? (Include both living and deceased; also include half

brothers and sisters) ....

(b) How many siblings has your spouse had, reckoned in same way? ....

(c) If your present marriage is not your first, give the number of siblings each previous spouse had ....

....

15. (*a*) Your religion (check) : Catholic ....; Protestant ....; Other ....; None ....

   (*b*) Your spouses's religion : Catholic ....; Protestant ....; Other ....; None ....

   (*c*) If this marriage is not your first, state religion of each previous spouse ....

   ....

16. The happiness of your marriage :

   (*a*) Do you and your spouse engage in outside interests together? (check) All of them ....; Most of them ....;

   Some of them ....; Very few of them ....; None of them ....

   (*b*) Do you ever regret your marriage? (check) Frequently ....; Occasionally ....; Rarely ....; Never ....

   (*c*) Have you ever seriously contemplated either separation or divorce? (check) Yes .... No..........

   (*d*) Can you truthfully say that your spouse never does or says anything that irritates or bores you in the slightest?

   (check) Completely true ....; Almost completely true ....; Questionable ....; Untrue ....

   (*e*) Can you truthfully say that when you have any unexpected leisure you always prefer to spend it with your spouse?

   (check) Completely true ....; Almost completely true ....; Questionable ....; Untrue ....

   (*f*) Everything considered, how happy has your marriage been? (check) Extraordinarily happy ....; Decidedly

   more happy than the average ....; Somewhat more happy than the average ....; About average ....; Per-

   haps a little less happy than the average ....; Definitely less happy than the average ....; Extremely un-

   happy ....

   (*g*) How well mated are you and your spouse from the strictly sexual point of view? (check) No two could be more

   perfectly mated sexually ....; Extremely well mated ....; Reasonably well ....; Not well ....; Very

   badly ....

Gifted Children Follow-up
Stanford University, 1950–51

Name of person filling out blank ....

(Name to be cut out on this line)

Age of person filling blank .... Date of filling out blank ....

Sex ....
M or F

Code number of blank ....

THE HAPPINESS OF YOUR MARRIAGE

Note: This blank to be filled out by the husbands and wives of gifted subjects. Please be objective and frank. The data will be regarded as highly confidential and will be used for statistical purposes only.

LEWIS M. TERMAN

(*a*) Do you and your spouse engage in outside interests together? (check) All of them .... ; Most of them .... ; Some of them .... ; Very few of them .... ; None of them ....

(*b*) Do you ever regret your marriage? (check) Frequently .... ; Occasionally .... ; Rarely .... ; Never ....

(*c*) Have you ever seriously contemplated either separation or divorce? Yes .... No ....

(*d*) Can you truthfully say that your spouse never does or says anything that irritates or bores you in the slightest? (check) Completely true .... ; Almost completely true .... ; Questionable .... ; Untrue ....

(*e*) Can you truthfully say that when you have any unexpected leisure you always prefer to spend it with your spouse? (check) Completely true .... ; Almost completely true .... ; Questionable .... ; Untrue ....

(*f*) Everything considered, how happy has your marriage been? (check) Extraordinarily happy .... ; Decidedly more happy than the average .... ; Somewhat more happy than the average .... ; About average .... ; Perhaps a little less happy than the average .... ; Definitely less happy than the average .... ; Extremely unhappy ....

(*g*) How well mated are you and your spouse from the strictly sexual point of view? (check) No two could be more perfectly mated sexually .... ; Extremely well mated .... ; Reasonably well .... ; Not well .... ; Very badly ....

GIFTED CHILDREN FOLLOW-UP
STANFORD UNIVERSITY, 1950–51   Date..

REPORT OF FIELD WORKER

Name of subject...

Name of field worker.. Date of
interview ....

Informants ....
(If not subject, give relationship to subject)

Address ....

1. Additional education or plans for further study.
2. (a) Occupation, (b) vocational and avocational interests, (c) recreation, (d) dynamics (drive, consistency of goal, satisfaction with goals).
3. Special abilities: Nature, degree of success.
4. Attitude of subject toward gifted study, own giftedness, school acceleration, etc., leading possibly into discussion of inferiority feelings (if any), feelings of adequacy and confidence, aspirations, etc.
5. Intellectuality of interests; impression of cultural level.
6. Health, nervous tendencies, emotional adjustment.
7. Marital status. Include education, occupation, and personality of spouse.
8. Family constellation. Include parents, offspring and others in household.
9. Home, neighborhood, other evidences of socio-economic status, including financial worries.
10. Special notes and comments. Total impression.

[Here followed two and one-half pages of ruled space for the field worker's report on the ten items above.]

RATINGS OF SUBJECT

Subject						Spouse				
1	2	3	4	5	1. Appearance	1	2	3	4	5
1	2	3	4	5	2. Attractiveness	1	2	3	4	5
1	2	3	4	5	3. Poise	1	2	3	4	5
1	2	3	4	5	4. Speech	1	2	3	4	5
1	2	3	4	5	5. Vanity	1	2	3	4	5
1	2	3	4	5	6. Alertness	1	2	3	4	5
1	2	3	4	5	7. Friendliness	1	2	3	4	5
1	2	3	4	5	8. Loquacity	1	2	3	4	5
1	2	3	4	5	9. Frankness	1	2	3	4	5
1	2	3	4	5	10. Attention	1	2	3	4	5
1	2	3	4	5	11. Curiosity	1	2	3	4	5
1	2	3	4	5	12. Originality	1	2	3	4	5

Gifted Children Follow-up
Stanford University, 1955

Date of filling out this blank ....

INFORMATION BLANK

Full name .... Age at last birthday ....
(Married women include maiden name)

Address .... Telephone...................................

1. Education : List any courses taken and degrees or certificates received since 1950 : ....

2. Occupation and earned income (for income report annual salary before income tax deductions). If self-employed (doctor, lawyer, business owner) give equivalent of salary, i.e., income less operating expenses.

Year	Profession, job, or position	Exact nature of work	Earned income per year
1952			
1953			
1954			

3. Approximate annual income, if any, of self and spouse from sources other than earnings ....

4. Any special honors, awards, offices held in clubs or organizations, biographical listings, etc. (specify) ....

5. Marital status (check) Single .... Married .... Widowed .... Separated .... Divorced ....

   Date of marriage .... If this, or previous marriage ended in divorce, give date(s) ....

6. About your spouse : Name .... His (her) age ....
   (Maiden name of wife)

   Highest grade or college year of schooling .... What school or college? ....

   Present occupation .... Present annual earned income ....

7. Offspring: (If any grandchildren please attach list with name, sex, and age of each)

Name	Sex	Date of birth	School grade	If not living, age at death	Cause of death

8. Other information regarding your spouse or children that you think would be of interest ....

9. Special accomplishments, honors, or misfortunes among near relatives (parents, brothers, sisters) since 1950. If any deceased, give date and cause of death ....

....

....

10. Your general physical health since 1950: (underline) very good .... good .... fair .... poor .... very poor .... Illnesses, accidents or surgery in recent years ....

....

After effects ....

11. Has there been any tendency toward nervousness, special anxieties, emotional difficulties, or nervous breakdown? .... Date and nature of such difficulties ....

....

....

How handled (medical or psychiatric help, hospitalization, etc.) Give details ....

....

Present condition (free from difficulty, improved, no change, worse, etc.) ....

....

12. Add any further significant information regarding yourself or your family which has not been covered above or for which the space provided was inadequate. Also list here any publications, patents, or other creative work since 1950.

....

....

....

....

....

....

....

....

....

....

....

....

....

....

....

....

....

....

INDEX

Abilities: compared with generality, 9, 16; degree of unevenness, 9, 16; early indications of, 8; versatility shown in occupations and avocations, 106, 107 ff.; *see also* Special abilities

Acceleration in school, *see* School acceleration

Achievement quotients: as related to amount of schooling, 9; compared with control group, 9; *see also* Educational histories

Achievement tests: 1921–22, 5, 8 f.; 1928, 17

Achievement to 1955:
—appraisal of: educational achievements, 71 f., 144; vocational achievements, 106, 145 ff.
—biographical listings: of men, 146 f.; of women, 145; increase in, 149 f.
—compared with generality, 145 f.
—illustrations of achievement: men, 83 ff., 146 f., 150 f.; women, 88 ff., 145
—outlook for future achievement, 149 ff.
—personality factors as determiners of achievement, 148 f.
—*See also* Success

Adjustment, general, 28, 35 ff., 148 f.; *see also* Mental health and general adjustment

Alcoholism, 21, 28, 143; as a mental disorder, 40; definition, 45; incidence compared with generality, 45 f.; ratings on use of liquor, 44 f.; sex difference in, 50 f.; subjects hospitalized, 40

American Men of Science: 145 f., 150; listings of men, 146; listings of women, 88, 145; listing as indication of achievement, 145 f.

Analogies test, 52, 55

Anderson, E. E., 153

Anderson, J. E., 153

Anthropometric measurements, 5, 6, 8, 34

Army Alpha Test, subjects selected by, 3

Avocational interests, 107 ff.; 117; as related to special abilities, 110 ff.; number of, 107 ff.; reading, 109; leading hobbies, 108 f.; as related to age, 107 f.; sex differences in, 108; *see also* Special abilities

Background of study: origin and purpose, vii ff.; plan of research, 1 ff.

Ballin, Marian, xiii

Bayley, Nancy, xii, 61, 153

Buchholtz, Shiela, xiii

Bunzel, B., 44, 153

Burks, Barbara S., 153

Cady, V. M., 11, 153

Careers of men and women, 73 ff.; *see also* Occupational status

Carnegie Corporation, v

Cattell, Jacques, 153

Character tests: scores of gifted and control subjects, 5, 11 f., 16

Ciocco, Antonio, 153

Columbia Foundation, v

Commonwealth Fund, v, viii, 1

Community service: 107, 115 f.; recognition and honors, 115

Composite portrait of typical gifted child, 1922, 15 f.; 1927–28, 18 f.

Concept Mastery scores: as related to Binet IQ, 57 ff., 143 f.; as related to education, 57 ff.; as related to marriage and divorce, 135; as related to mental health and general adjustment, 49 f.; as related to radicalism-conservatism, 125 f., 131; method of equating Forms A and T scores, 61 f.; of spouses compared with college groups, 138; of subjects and spouses, 56 f.; of various groups tested, 60; sex differences, 57; test-retest comparisons, 61 ff., 142 ff.; *see also* Concept Mastery test; Intelligence

Concept Mastery test, 23, 24, 25, 27, 52 ff., as affected by schooling, 53; correlation between Forms A and T, 55 f.; description of, 52 ff.; Form A compared with Form T, 54 f.; groups tested, 59 f.; normative data, 59 ff.; reliability and validity, 54 f.; score distributions of subjects and spouses, 56 f.; *see also* Concept Mastery scores

Co-operation of subjects, xi, 20, 21, 25 f.

Cox, Catharine M., xi, 153

Crime and delinquency, 28, 46, 51

Data on Rate of Reproduction blank, 23, 24, 27, 172 ff.

Davis, Kingsley, 153

Delinquency, 21; *see also* Crime and delinquency

Directory of American Scholars, 145, 146; listings of men, 146 f.; listings of women, 145; listing as indication of achievement, 145 f.

Divorce: incidence of, 133 ff.; as related to education, 133 f.; as related to general adjustment rating, 135; as related to intelligence, 135; as related to scores on Marital Happiness Test, 135 f.

Doyle, Babette, xiii

Dublin, L. I., 28, 31, 32, 44, 139, 153

DuBois, Cornelius, 153

Early development, *see* Physical history

Education of gifted children, contributions to, xii, 72

Educational histories:
—acceleration, 8, 21, 22, 72
—achievement quotients, 9
—achievement tests, 1922, 8 f.; 1928, 17
—age at completing high school, 72
—age at entering school, 8
—age at learning to read, 8
—amount of education compared with parents' attitude toward college attendance, 71
—amount of education, 64 ff.
—amount of education as related to: divorce, 133 f.; education of parents, 71; income (men), 94 f.; income

(women), 98 f.; intelligence, 57 ff.; marriage, 133; radicalism-conservatism, 122
—college grades, 67 f.
—college graduation, 64 ff.
—college graduation compared with generality, 65
—educational record appraised, 71 f., 144
—failures in college, 68
—graduate study and degrees, 65 ff.
—graduate study compared with generality, 66 f.
—graduation honors, 67 f.
—major fields of study, 68
—school achievement, 1928, 18
—schooling of parents, 6, 71
—subject's opinion on amount of schooling, 69
—subject's report on parents' attitude toward schooling, 70 f.
—*See also* School acceleration

Family background of subjects, 5 f.; *see also* Parents

Fertility: 139 f.; of group to 1945, 22; as related to age of gifted women, 140; as related to age of wives of gifted men, 140; *see also* Offspring

Field workers, xii, 56; reports of, 17, 20 f., 26, 45, 177 f.

Follow-ups:
—1927–28, 17 ff.; data obtained, 17 f.; summary of findings, 18 f.
—1936, 19 f.
—1939–40, 20 ff.; data obtained, 21; summary of findings, 21 f.
—1945, 21; summary of findings, 21 f.
—1950–52, 23 ff.; data obtained, 23 f.; information schedules, 23
—1955, 23 ff.

Fund for the Advancement of Education, v

Galton, Francis, 22

Gebhard, P. H., 154

General adjustment, *see* Mental health and general adjustment

General Information Blank, 1950: 23, 24, 27, 44, 107, 114, 119, 160 ff.

Goldhamer, H., 42, 43, 153
Goodenough, Florence L, xii, 153
Happiness of Your Marriage Blank, 24, 27, 135 f., 176
Havemann, Ernest, 79, 80, 98, 117, 133, 134, 154
Health history, 6 ff.; 1928, 18; *see also* Physical health
Height: reported by subjects, 1940, 34; *see also* Anthropometric measurements
Henry, A. F., 32, 154
Hirning, Alma L., 154
Hirning, J. L., 154
Hobbies, *see* Avocational interests
Homosexuality, 21, 47 ff., 51, 143

Income, earned:
—of men, 93 ff.; as related to age, 93 f.; as related to education, 94 f.; as related to opinion on extent abilities lived up to, 104; as related to radicalism-conservatism, 124 f.; as related to vocational satisfaction, 104; as a source of satisfaction, 105; by occupation, 95 ff.; comparison with generality, 97 f.
—of women, 98 f.; as related to education, 98 f.; as related to radicalism-conservatism, 125; as a source of satisfaction, 105; by occupation, 99
Income: total family, 100 f.; compared with generality, 100 f.; as related to radicalism-conservatism, 125
Information About Child blank (offspring), 23, 24
Information Blank, 1955: 24, 27 ff.
Information obtained:
—1921–23, 4 ff.
—1927–28, 17 f.
—1936, 19 f.
—1939–40, 21
—1945, 21
—1950–52, 23 ff.
—1955, 24 ff.
Information schedules reproduced: 157 ff.
—Data on Rate of Reproduction, 172 ff.
—General Information (1950), 160 ff.
—The Happiness of Your Marriage, 176
—Information Blank (1955), 179 f.

—Report of Field Worker, 177 f.
—Supplementary Biographical Data, 164 ff.
Intellectual abilities: subjects opinion on extent lived up to, 104 f.
Intellectual status:
—as of 1940, 22, 54; as of 1950–52, 52 ff.
—compared with generality, 54, 60 f.
—estimate of changes in: to 1928, 18; to 1940, 54; to 1952, 61 ff.
—maintenance of intellectual superiority, xii, 61 ff.; 143 f.
Intelligence:
—and amount of education, 57 ff.
—correlation with desirable traits, 142
—maintenance of, 61 ff.; 143 f.
—of offspring, 141 f.
—as related to general adjustment, 49 f.
—as related to marriage, 135
—as related to radicalism-conservatism, 125 f.
—*See also* Concept Mastery scores; Intellectual Status; Stanford-Binet test
Intelligence tests:
—1921–22, 2 f.
—1927–28, 17 f.
—1939–40, 52 ff.
—1950–52, 52 ff.
Interest Blank, for subjects:
—1921–22, 5, 9 f.
—1928, 17
Interests: avocational, 107 ff.; reading interests, 109, 117; versatility of 107 ff.
Interests in childhood, 9 ff.; in plays and games, 10; reading interests, 11; scholastic interests, 9; vocational interests, 10; *see also* Play interests
IQ: distribution of IQ's of offspring, 141 f.; of subjects when selected, 2 f.; predictive value of, 144; subjects with IQ 170 or above, 21, 125 f.; *see also* Intelligence

Jellinek, E. M., 46, 154
Jensen, Dortha W., 153
Juvenilia, literary, 18 f.

Keller, Mark, 46, 154
Kinsey, A. C., 48, 51, 154

Labor organizations: activity in, 114; membership in, 113 f.
Landis, Carney, 154
Lange-Eichbaum, Wilhelm, 148, 154
Lehman, H. C., 149, 154
Lotka, A. J., 153

McNemar, Olga W., xii
McNemar, Quinn, viii, xiii, 4, 154
Malzberg, Benjamin, 42 f., 154
Marital aptitude test, 22
Marital happiness test, 22, 23, 24, 47, 135 f.
Marital selection:
—age difference between spouses, 137 f.
—education of spouses, 58, 137
—intelligence of spouses: 58, 138; compared with college groups, 138 f.
—intermarriages in group, 139
—occupational status of spouses, 137 f.
—See also Marriage
Marriage:
—age at, 132
—incidence to 1945, 22
—incidence by age to 1955, 132 ff.
—incidence by amount of education, 133
—marital happiness: scores on happiness test, 135 f.; rating on happiness of marriage, 136; compared with generality, 136
—as related to intelligence, 135
—See also Divorce; Fertility; Marital selection
Marsden Foundation, v
Marshall, A., 42, 43, 153
Marshall, Helen, xii
Martin, C. E., 154
Medical examinations, 4 f., 7 f., 34
Memberships in clubs and organizations, 107, 112, 117; kinds of, 113; number of, 112
Mental adjustment, see Mental health and general adjustment
Mental deficiency, among offspring of subjects, 142
Mental disease: 21, 28
—age at hospitalization, 41
—definition of, 41 f.
—diagnosis of illness, 40
—expectancy of, 42 f.
—incidence compared with generality, 41 ff.

—length of hospitalization, 39 f.
—number hospitalized, 37 ff.
—recovery from, 38 ff.
—subjects hospitalized, 37 ff.
—See also Alcoholism; Suicide
Mental health and general adjustment, xii, 28, 35 ff., 143
—basis for ratings, 35
—definition of ratings, 35 f.
—incidence of mental disease, 21, 37 ff.
—mental disease defined, 41 f.
—ratings on mental health and general adjustment, 36 ff.
—as related to education, 49 f.
—as related to intelligence, 49 f.
—as related to marital status, 135
—as related to radicalism-conservatism, 126 f.
—See also Alcoholism; Delinquency; Homosexuality; Mental disease, Suicide
Merrill, Maud A., 154
Minnesota Occupational Scale, 73, 80, 81, 87, 137
Mortality, 24, 28 ff.; age at death, 29; causes of, 29 f.; compared with generality, 29; education of deceased subjects, 32 f.; incidence of, 29; IQ of deceased subjects, 32; mental illness among deceased subjects, 38 ff.; occupation of deceased subjects, 33; suicides, 30 ff.
Murphy, J. V., 153

National Academy of Sciences: elections to, 146, 150
National Intelligence Test, 2, 3, 4; subjects selected by, 2 f.
National Research Council, v
Nervous disorders, see Mental health and general adjustment
Norris, Dorothy, 155

Occupational classification, see Minnesota Occupational Scale, also Occupational status
Occupational status:
—appraisal of, 106, 144 ff.
—gifted men: as of 1955, 73 ff.; breakdown of occupational groups, 74 ff.; classification of occupations, 73 f.; compared with generality of male

college graduates, 79 ff.; illustrations of occupations, 77 ff.; occupational changes between 1940 and 1955, 81 ff.; as related to income, 95 f.; as related to radicalism-conservatism, 123

—gifted women: as of 1955, 85 ff.; according to education, 86; according to marital status, 85 f.; breakdown of occupational groups, 87 f.; classification of occupations, 87; illustrations of careers, 88 ff.; as related to income, 99 f.; as related to opinion on how well intellectual abilities lived up to, 104 f.; as related to radicalism-conservatism, 124

—*See also* Achievement to 1955; Vocation

Occupations, childhood preferences, 10; *see also* Occupational status

Oden, Melita H., vii, 61, 153, 155

Offspring: incidence of twins, 140; intelligence of, 141 f.; number of, 139; n u m b e r according to education, 140 f.; number of adoptions, 142; number of deaths, 140; number of grandchildren, 142; sex ratio, 139 f.; *see also* Fertility

Page, J. D., 154

Parents: education of, 6, 71; home library, 6; income of, 6; listings in *Who's Who,* 6; occupations of fathers, 5 f.

Patents, number awarded to 1955, 88, 145, 147

Phi Beta Kappa, 67 f., 144

Physical defects, 7 f., 34 f.

Physical health, 6 ff., 28 ff., 143; self-ratings, 33 ff.

Physical history: early development, 7 f.; birth history, 7; *see also* Health history

Play information, test of, 10 f.

Play interests, 16; as measures of interest maturity, 10; as measures of masculinity-femininity, 10; as measures of sociability, 10; test of, 5, 10

Political and social attitudes:

—party preferences, 127 f.

—political activities, 129 f.; offices held, 129 f.

—self-ratings on radicalism-conservatism, 118 ff.; as related to age, 121 f.; as related to education, 122; as related to general adjustment, 126 f.; as related to income, 124 f.; as related to intelligence, 125 f.; as related to occupation, 123 f.; as related to political preferences, 128; comparison of 1940 and 1950 ratings, 120 f.; sex differences in, 120

—summarized, 130 f.

—voting habits, 129

Pomeroy, W. B., 154

Pressey Group Test, 4

Psychological Corporation, 54

Public offices: election to public office, 129 f.; appointive offices held, 129 f.; *see also* Political and social attitudes

Publications, number and kinds: for men, 110, 147 f.; for women, 110, 145

Racial origin, 5

Radicalism-conservatism, *see Political* and social attitudes

Raubenheimer, A. S., 11, 12, 154

Reading, 5; age at learning, 8; reading interests as adults, 109, 117; reading interests in childhood, 11

Religion: affiliations, 116; attitudes toward, 116 f., 118; religious training, 116; sex differences in membership, 116 f.

Reproduction, rate of, *see* Fertility

Rockefeller Public Service Award, 130

School acceleration, 8, 21, 22, 72; and graduate study, 72; as a method of providing for the gifted, 72; progress quotients, 8 f.

Schooling, *see* Educational histories

Scottish investigations, sex differences in variability, 4

Sears, Robert, xiii

Selection of subjects: procedures used, 2 ff.; tests used, 2 ff.

Semelman, Barbara B., 32, 154

Service activities, 107, 114 ff., 118; contributions to community life, 115; memberships in community welfare groups, 115; recognitions and honors for, 115 f.

Sex adjustments, 22, 47 ff.

Sex differences: in alcoholism, 46, 50 f.; in amount of schooling, 69; in avocational interests, 108; in church membership, 116 f.; in Concept Mastery scores, 57; in incidence of crime and delinquency, 51; in incidence of homosexuality, 51; in incidence of mental disease, 50; in IQ, 3 f.; in IQ of offspring, 141 f.; in mortality rate, 34; in play interests, 10; in radicalism-conservatism, 120; in ratings on health, 34; in relationship of general adjustment to education and intelligence test scores, 51; in suicide rate, 32, 50; in vocational goals, 73, 106, 145

Sex problems, 28, 46 ff.

Sex ratio: among subjects, 4; of offspring, 139 f.

Shea, Alice Leahy, xii

Short, J. F., Jr., 32, 154

Siblings, included in gifted group, 2, 6

Sigma Xi, 67 f., 144

Social attitudes, see Political and social attitudes

Social origin, 5

Spearman, Charles, 52

Special abilities: 110 ff., 150 f.; in art, 111; in dramatics, 111; in music, 110 f.; illustrations of special ability, 111 f.; as related to vocations and avocations, 110 ff.; writing as an avocation, 110

Spiegelman, Mortimer, 153

Spouses, of gifted, see Marital selection

Stanford Achievement Test, 1922, 8 f.

Stanford-Binet IQ's: changes to 1928, 18; compared with Concept Mastery scores, 57 f.; distributions for offspring, 141 f.; of subjects, 3; see also Intelligence

Stanford-Binet test, viii, 1 ff., 23, 24, 27, 61, 141 f.; subjects selected by, 3; see also Stanford-Binet IQ's

Strong, E. K., 154

Strong Vocational Interest Blank, 21;

Study: in adult education classes, 67; in study groups, 109, 117; through independent reading, 109, 117

Subjects: basis for selection, 2 f.; cooperation of, xi f., 20, 21, 25 f.; deceased subjects, 24, 29 ff., 32; effects of inclusion in group, 20; the group as of 1928, 17; the group as of 1940, 20 f.; the group as of 1952, 24; the group as of 1955, 24; intermarriages among, 139; IQ's of subjects, 3; method of selection, 2 f.; number lost, 24, 25; number of, 3; sex ratio among, 4; siblings and cousins in group, 6; proportion living in California, 23, 32

Success: differences found in comparison of most and least successful men, 148 f.; measures of, 145, 148 ff., 151 f.; subjects' opinion of what constitutes success, 151 f.; see also Achievement to 1955

Suicide, 30 ff.; and mental disorder, 43 f.; incidence compared with generality, 30 ff.; sex differences in, 32, 50

Sullivan, Ellen, xii

Sumption, M. R., 155

Supplementary Biographical Data blank, 24, 27, 47, 64, 68, 101, 116, 136, 151, 164 ff.

Sward, Keith, 53, 154

Synonym-antonym test, 52, 54 f.

Terman Group Test, 2 f.; subjects selected by, 2 f.

Terman, Frederick E., viii

Terman, L. M., vii ff., xi ff., 4, 19, 146, 153, 154, 155

Thomas Welton Stanford Fund, v

Thorndike CAVD test, 4

Trait ratings by parents and teachers, 12 ff.; gifted and control groups compared, 13 ff.

U.S. Census reports: current population reports, 100, 155; statistical abstract, 97, 155

Vocation: as a source of satisfaction, 105; subjects' feelings about, 101; vocational achievements, 106; vocational satisfaction as related to income, 104; as related to occupation, 102 ff.; see also Occupational status

Voting habits, see Political and social attitudes

Wallin, Paul, 154
Wechsler-Bellevue test, 61
Weight: reported by subjects, 1940, 34;
  see also Anthropometric measurements
West, Patricia Salter, 80, 98, 117, 133, 134, 154
Whittier Scale for Grading Home Conditions, 6
Who's Who in America, 145, 146, 147, 150, 155; listings of men, 146 f.; listings of women, 145; listing as indication of achievement, 145 ff.; parents listed in, 6
Wolfle, Dael, 65, 66, 71, 155
Woodbury, R. M., 7, 155
World Health Organization, definition of alcoholism, 45, 46, 50
Writing: as an avocation, 110; see also Avocational interests

Yearbook of American Churches, 155